I0889719

THE LIBRARY ON DISPLAY

12

In memoriam
Rafael Tovar y de Teresa

Catalogue of the exhibition held in the Biblioteca Medicea Laurenziana
from 2 November 2023 to 2 February 2024.

Scientific collaborators

Getty Research Institute

With the contribution of

In collaboration with
Biblioteca Nazionale Centrale, Florence
Le Gallerie degli Uffizi, Florence
Museo Archeologico di Firenze/Direzione Regionale dei Musei della Toscana
Museo delle Civiltà, Rome

Concept and Scientific Project
Ruby E. Villarreal V.

Scientific committee
Francesca Gallori, Lorenzo Hofstetter, Silvia Scipioni, Ruby E. Villarreal V., Juan Villoro Ruiz

Exhibition Layout
Ruby E. Villarreal V., Claudio Chiarusi, Cosimo Frezzolini, Roberto Maltagliati

General Coordination
Biblioteca Medicea Laurenziana:
Francesca Gallori, Eugenia Antonucci
'Entre dos Mundos' cultural association:
Maurizio Catalano
Honorary Consulate of Mexico in Florence:
Daniela Konyk

Multimedia Presentation
Simone Falteri, Mario Wong, Secretaria de Turismo de Tabasco, Ruby E. Villarreal V.

Press Office
Biblioteca Medicea Laurenziana

Exhibition Graphics
Maria Gabriela Ramirez

Exhibition Installation
Stampa in Stampa

Ticket Office and Reception
Opera d'Arte

BIBLIOTECA
MEDICEA
LAURENZIANA

RUBY E. VILLARREAL V.

cacao entre dos mundos

the history of chocolate and the clash (and meeting) of civilisations

with contributions by
FILIPPO CAMEROTA
LORENZO HOFSTETTER

Mandragora

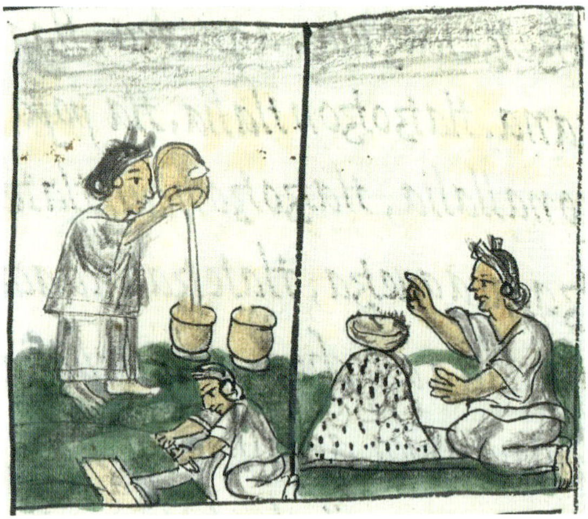

Women Making Chocolate. Florence, Biblioteca Medicea Laurenziana, Mediceo Palatino 220, Mexico, 16th century (1577), fol. 71v, detail.

PREFACE

Preserving and circulating documentary heritage helps toward a greater understanding and dialogue between peoples, fostering peace, human rights and dignity: this is the driving principle behind UNESCO's 'Memory of the World' programme and its International Register of the Memory of the World. Since 2015, this register has included the so-called 'Florentine Codex', namely, the *Historia general de las cosas de Nueva España*, the extraordinary encyclopaedia of Aztec civilisation conceived by Bernardino de Sahagún and preserved in our library.

The Biblioteca Medicea Laurenziana has always shared this conviction and so was delighted by the idea of an exhibition dedicated to cocoa – one of the most precious treasures from the New World – as described and narrated in Brother Bernardino de Sahagún's work. According to Aztec mythology, cocoa has a divine origin, closely connected to the god Quetzalcóatl. The exhibition begins here, enhanced by important loans, and for this we thank the museum and library directors. It then moves to the successful reception of chocolate in 17th and 18th century Europe and then in our lives.

I wish to give my heartfelt thanks to Cosimo di Nocera, Honorary Consul of Mexico in Florence, and architect Ruby Villarreal, Cultural Attaché of the Consulate who, together with the 'Entre dos Mundos' cultural association, conceived and proposed the exhibition and organised it in collaboration with the Biblioteca Medicea Laurenziana; I am also grateful for the significant scientific support of the Getty Research Institute and the Galileo Museum of the History of Science in Florence. Finally, a heartfelt thanks goes to the Region of Tuscany and its President, Eugenio Giani, and the Fondazione CR Firenze for their generous support of the enterprise.

Francesca Gallori
Director of the Biblioteca Medicea Laurenziana

PRESENTATION

Humani nihil a me alienum puto
Publius Terentius Afer,
Heautontimorùmenos (165 BC), l. 77

The first time I had the privilege of admiring the Florentine Codex in person was thanks to Rafael Tovar y de Teresa during his first official visit to Florence, after he had been appointed Mexican Ambassador to Italy. Don Rafael's excitement over seeing Brother Bernardino de Sahagún's original manuscript pages in the small room where we were allowed to look at them in private was enormous. And for me, it was love at first sight. It was also the first time I had admired the interior of the marvellous Biblioteca Medicea Laurenziana, designed by the divine Michelangelo at the request of the Medici pope, Clement VII. In that moment I realised that the richest and most profound connection between Mexico and Italy is enshrined in this marvellous treasure chest of Renaissance architecture in Florence. The year was 2003.

Some time later, I attended the conference organised by the Biblioteca Medicea Laurenziana on the *Historia general de las cosas de Nueva España* – the occasion was the 2007 European Heritage Days. I was surprised by the questions from the audience at the end of the speeches, since among the various topics discussed – which ranged from human sacrifices and anthropophagy to the Aztec calendar and the pyramids – it was the subject of chocolate that caught the audience's attention. In fact, the original recipe for chocolate is transcribed in the Florentine Codex, dated 1577, along with its many variations. This was the moment when the idea of 'Entre dos Mundos' came into being, a project that would research and explore how the Old and New Worlds influenced each other, leaving indelible traces in their history and culture. Todorov was right when he repeatedly affirmed that 'The past is fruitful not when it serves to nourish resentment or triumphalism, but when its bitter taste compels us to transform ourselves' (*L'Homme dépaysé*, Paris, 1996, p. 71).

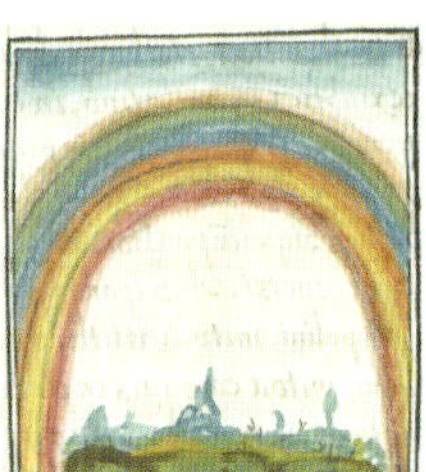

'Ayaucoçamalotl' (rainbow). Florence, Biblioteca Medicea Laurenziana, Mediceo Palatino 219, Mexico, 16th century (1577), fol. 238r, detail.

Sahagún's books led me to other extraordinary texts, which were deeply inspiring for me. The first was *Colors Between Two Worlds. The Florentine Codex of Bernardino de Sahagún*, published by The Harvard University Center for Italian Renaissance Studies. Its evocative title came from Diana Magaloni Kerpel, former director of the Museo Nacional de Antropología, and her intense description of the Florentine Codex: 'The result, a truly unique masterpiece … is also a history of the conflict, negotiation and dialogue between two different worlds … it is a carefully and skilfully crafted work of art whose material qualities and historic context bear the indelible marks of an epic battle for survival' (Ida Giovanna Rao, Diana Magaloni, Alessandra Pecci, *Il mondo degli Aztechi nel Codice Fiorentino*, Florence, 2007, pp. 12–13). The image Ayaucoçamalotl, from the Sahagún Codex, decorates the back cover of the volume. It is a colourful rainbow, which literally acts as a symbol uniting two opposites. A fascinating emblem for the uniting of antipodes. The 'Entre dos Mundos' cultural association project took its name and logo from *Colors Between Two Worlds*. The second book that inspired me was *La conquête de l'Amérique. La question de l'autre*, by Tzvetan Todorov – where 'the Other' is 'the Different one' –, a text that encourages us to reflect on the relationship between individuals belonging to different cultures and social groups starting with historical memory: 'I am writing this book to prevent this story and a thousand others like it from being forgotten. I believe in the necessity of "seeking the truth" and in the obligation of making it known … My hope is … that we remember what can happen if we do not succeed in discovering the other' (Paris 1982, Eng. trans. New York 1984, p. 247). Finally, the pivotal book, the primary source for the exhibition and educational project 'Cacao Entre Dos Mundos', I found during a wonderful visit to the Museo de Anthropologia in Mexico City: the Spanish edition of the most thorough publication on the history of chocolate, *The True History of Chocolate* by Sophie D. Coe and Michael D. Coe, anthropologists at Yale University, revised in Rome by Michael after the untimely death of his Sophie. The authors use the original sources on the history of cocoa because 'much food writing about chocolate's past falls into the category of Voltaire's "accepted fiction".' (London 1996, p. 12).

The 'Cacao Entre Dos Mundos' project is being realised on the occasion of the 500th anniversary of the start of the Spiritual Conquest that took the friar Bernardino de Sahagún to Mexico. The first venue is the majestic Biblioteca Medicea Laurenziana. Against the backdrop of the resplendent green of the quetzal, maps and archaeological, historical, artistic and artisan artefacts tell the story of chocolate, from the Mesoamerica of the Olmecs to the 'bean to bar' movement, a beacon of hope in respect to 'the Other'. The narrative guide is provided by the twelve books of the *Historia general de las cosas de Nueva España* by Sahagún, who, as Miguel León-Portilla reminds us, with 'his humanism as a Renaissance Spaniard, he made the saying that no human reality can be foreign become true' (*El México Antiguo en la historia universal*, Mexico, 2015, p. 77).

Ruby E. Villareal V.
President of the 'Entre dos Mundos' cultural association

Acknowledgements
The collective creation common to the indigenous world and intrinsic in Sahagún's work also applies to the exhibition 'Cacao Entre Dos Mundos', which is the result of a great deal of collective work: Antonio Aimi, Soraya Al azzam, Eugenia Antonucci, Francesca M. Anzelmo, Roberta Avena, Nicoletta Bardossi, Ilaria Bartocci, Giulia Basilissi, Diana Beltrán, Filippo Camerota, Roberto Caraceni, Azzurra Carmosino, Laura Carrillo, Stefano Casciu, Maurizio Catalano, Guido Catani, Rebeca Chavez, Claudio Chiarusi, Martin Christy, Mario Curia, Eduardo de la Cruz, Cosimo Di Nocera, Elisa Díaz, Davide Domenici, Lourdes Echavarria, Simone Falteri, Davide Ferrero, Cosimo Frezzolini, Elyn Fuentes, Francesca Gallori, Eugenio Giani, Marcela Gómez, Gabriele Gori, Carla Guarducci, Lorenzo Hofstetter, Mario Iozzo, Daniela Konyk, Elena Landini, Donatella Lippi, Roberto Maltagliati, Roberta Masini, Ottaviano de' Medici, Ángel Méndez, Monica Meschini, Andrea Micozzi, Eduardo Matos Moctezuma, Massimo Montanari, Vincenzo Muscolo, José Antonio Nieves, Alejandra Ortiz, Cinzia Ortri, Maurizio Palatresi, Monica Patiño, Mauro Pellegrino, Claudio Pistocchi, Gabriela Ramirez, René Renteria, Kim Richter, Pierpaolo Ruta, Valentina Saetti, Maria Salemi, Marco Salucci, Enrico Sassonia, Silvia Scipioni, Martha Sepulveda, Paolo Staccoli, Antonietta Troiano, Paola Vannucchi, Nicolò Vellino, Danielo Vestri, Andrea Viliani, Elisa Villarreal, Juan Villoro, and Mario Wong.

Thank you for the invaluable support: Rogelio Hernández, Humberto Tijerina.

Thank you for the special emotional support: Babbo Alfonso, Jorge Guerra, Raquenel Hernández, Melisa Martínez, Lara Pasquarelli, Luca Pasquarelli, Mamma Ruby, Cynthia Salas, Jorge Tijerina, Lupita Tijerina, Elisa Villarreal.

Thank you for your immense heart: Antica Dolceria Bonajuto, Casa Nahuatl, La Cucaracha, Gelateria della Passera, Istituto Lorenzo de' Medici, Nicolò Vellino, Staccoli, Vestri Cioccolato.

Thank you above and beyond for enriching the project before getting started: Raquenel Cázares, Edmundo Escamilla, Yuri de Gortari, Zoila Vargas.

Authors of the texts
[L.H.] Lorenzo Hofstetter
[R.V.] Ruby E. Villarreal V.

INTRODUCTION

On the occasion of their 500th anniversary, we recall two conceptually distant yet converging events: the start of Michelangelo Buonarroti's commission from Pope Clement VII to design the Biblioteca Medicea Laurenziana, and the beginning of the 'Spiritual Conquest' of the New World with the arrival of the first religious brothers of the mendicant orders in America in 1523. These events determined the genesis and fate of an extraordinary American manuscript in twelve books, produced by a Franciscan friar, preserved in the Biblioteca Medicea Laurenziana. Considered the first anthropological encyclopaedia of all time, within it the story of cocoa is recounted and illustrated, prompting us to reflect, starting with the historical approach, on those still sensitive aspects of the world of chocolate that are among the sustainable development goals of Agenda 2030: responsible consumption and production, environmental sustainability, social justice, eradication of poverty and an end to child labour and modern slavery.

Scientific and exhibition project

'Cacao Entre Dos Mundos' is a journey through time that begins with the collision of two worlds, from the moment when Admiral Cristoforo Colombo sighted a land never before seen by medieval westerners – 'Mundus Novus', as the Florentine navigator Amerigo Vespucci first called it. This unleashed a dramatic and historic clash of civilisations, culminating in the Spanish colonisation, following the tragic fall of the Aztec empire caused by the *conquistadores* led by Hernán Cortés.

The journey continues in the wake of the Spiritual Conquest brought about by expeditions of missionaries of the mendicant orders to the New World. Among them was 'our' Bernardino de Sahagún, a Franciscan, considered the founder of modern anthropology thanks to his *magnum opus* on the Aztec people, *Historia general de las cosas de Nueva España*, where the recipe for chocolate is transcribed for the first time.

We then retrace the arrival of cocoa in Europe following the so-called 'Columbian Exchange', together with the Medici family's involvement in spreading the unusual beverage; we find scientific information on the organoleptic properties and the nutritional and therapeutic qualities of the precious fruit, the transformation of cocoa into chocolate, and the evolution of its processing and consumption over the centuries. Finally, the cocoa market is analysed, based on globalised production that overexploits natural resources, makes extensive use of child labour and is not above abusing women. The exhibition concludes with an analysis of projects that positively impact environmental and social sustainability in the cocoa sector, emphasising the vital importance of historical memory as a foundation for building a better future.

The documentary heritage is revealed by enhancing the scientific content through innovative approaches, such as storytelling for the captions and the use of new technologies within the traditional layout. The exhibition consists of valuable manuscripts, original and digital maps, Mesoamerican archaeological finds, indigenous artisan artefacts, mediaeval coins, Renaissance portraits, Baroque porcelain, valuable reproductions of historical relics and multimedia exhibits.

The exhibition space is divided into six stations, five of which correspond to the relevant thematic references described in the twelve books of the Florentine Codex. These provide input, outlining each salient phase in the history of cocoa, while the last and sixth station is dedicated to the 'Memory of the World'. [R.V.]

At the meeting between the Old and New Worlds – where our journey begins – there could only be an emblem that is as imposing as it is undeniably visually striking: the *Piedra del Sol*. Commissioned by Moctezuma II, it encloses an entire imaginative, ritual and scientific universe within an olivine basalt circle almost 4 metres in diameter and 24 tonnes in weight. Carved using only flint tools, the stone was originally located in the main square of the Aztec capital Tenochtitlán (today, Mexico City). Most likely placed in the Great Pyramid dedicated to Tlaloc, god of rain, and Huitzilopochtli, personification of the Sun and god of war, it consists of a series of concentric circles centred on the cosmology and theology of the Mexica people. Starting from the outside, the fifth circle depicts two impressive snakes, from whose mouths peep the deities Xiuhtecuhtili and Tonatiuh, corresponding to the skies of night and day respectively. The coils, composed of flames, cross at a point bearing the date of completion of the sculpture: 'matlactli omey acatl' ('thirteen reeds'), corresponding to 1479. The fourth and third circles are marked by four large rays of sunlight culminating in as many corners, superimposed on a band of symbolic elements recalling thorns, eagle feathers and drops of human blood, the food that the Sun needs to rise each morning. The second circle is the key to reading the calendar, subdivided into eighteen months of twenty days each, each of which is represented here with its own symbol, including *miquiztli*/death, *cozcacuauhtli*/vulture and *ocēlotl*/jaguar. The length of the year, calculated from extensive astrological knowledge also common among other Mesoamerican peoples, was 365 days (366 in leap years). The first circle depicts the succession of the four

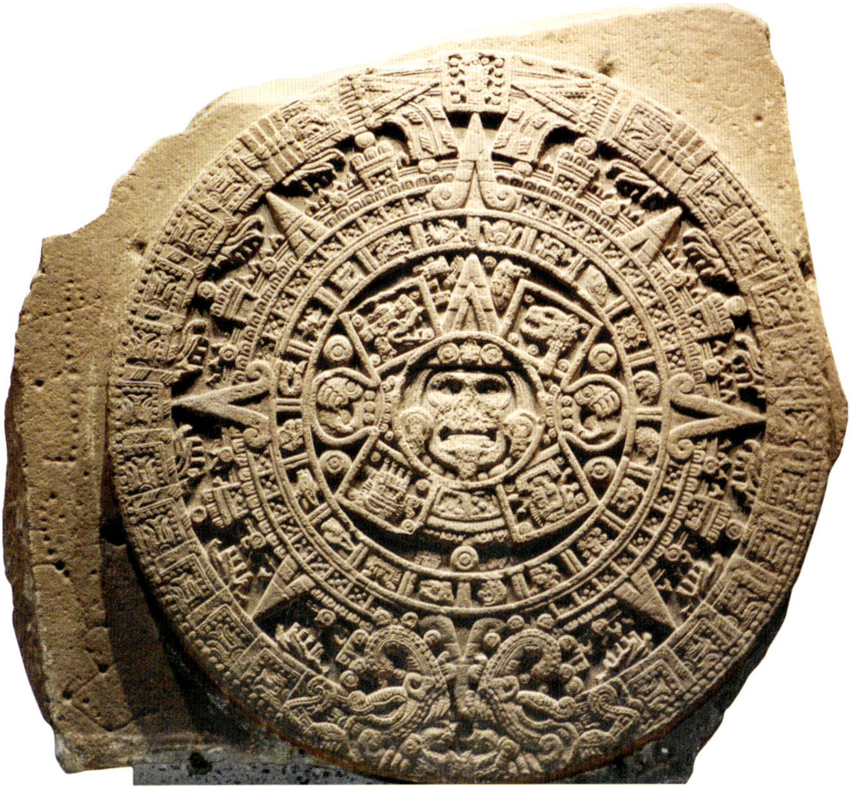

Mexica art, *Piedra del Sol*, Mexico City, 15th century, olivine basalt, ø 3.6 m, Mexico City, Museo Nacional de Antropología.

generating Suns of the World: the Jaguar (*nahui ocelot*), the Wind (*nahui ehēcatl*), the Rain of Fire (*nahui quiàhuitl*) and Water (*nahui atl*). In the centre is the Fifth Sun, Tonatiuh (Ollin Tonatiuh), whose cycle corresponds to the present era. [L.H.]

CATALOGUE

I. DIVINE

Book I, Mexica divinities
'que se trata de los dioses que adoraban los naturales desta tierra que es la Nueva España'

Book II, Ceremonies, festivities, rites, sacrifices
'que trata del calendario, fiestas y cerimonias, sacrificios y solenidades que estos naturales désta Nueva España hacían a honra de sus dioses'

Book III, Principle of the gods
'del principio que tuvieron los dioses'

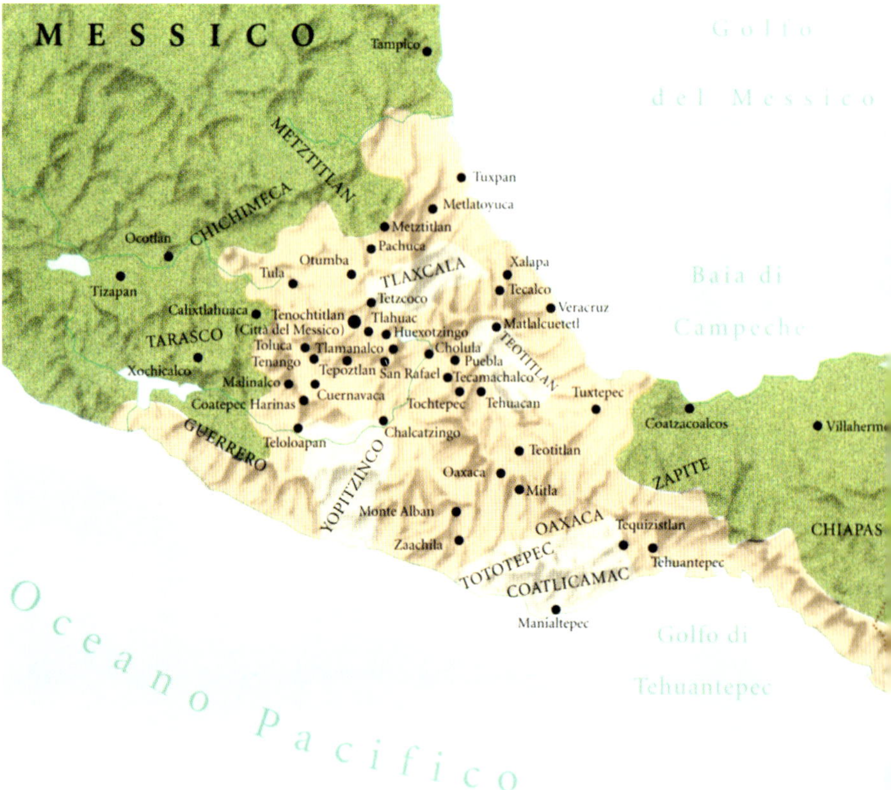

MESOAMERICA

Chocolate, one of the most exquisite gifts for the human palate, is made from the seeds of the fruit of the cacao tree which flourished in the lands of Mesoamerica. There are numerous myths and legends surrounding cacao, originating from the Native American cosmovision and the Mesoamerican deities. The expression 'Mesoamerica', does not merely refer to a geographical area, but encapsulates a precise cultural context in a vast area that encompassed the present-day Mexico, Guatemala, El Salvador, Belize, Honduras and Nicaragua and Costa Rica, before the arrival of the Europeans. The appearance of Mesoamerican civilisations dates back to 7000 BC, based on the evidence of agricultural activity that has come down to us from the Holocene epoch. The main Mesoamerican cultures, which are the most widely studied today, were the Olmec, Maya, Mixtec, Zapotec, Teotihuacan, Toltec and Mexica, also known as Aztec.

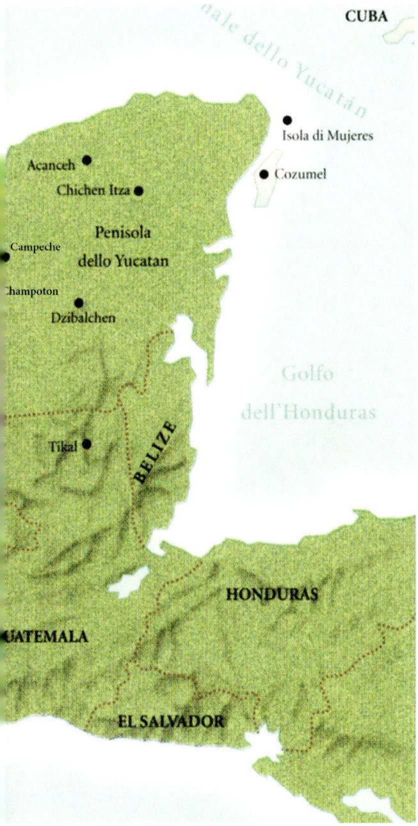

Rather than being a Tower of Babel of different languages and peoples, Mesoamerica was characterised by a marked uniformity in its socio-political systems, which were structured as confederations of city-states with the most powerful urban centres at the apex, notable for the presence of sophisticated and monumental architectural works such as the pyramids. They practised a polytheistic religion, used human sacrifices for ritual purposes and used images of animals to represent certain gods. These civilisations were characterised by a common lithic technology; the absence of any metallurgical activity; the use of hieroglyphic writing and identical calendars (one civil and one religious). Essentially agricultural societies, they developed the cultivation of maize, beans, tomatoes, avocados, vanilla, pumpkin and chilli peppers and cacao. In fact, there is archaeological evidence, dating back 3750 years, of the consumption of chocolate by the Olmecs at the sacred site of Cerro Manatí (Veracruz, Mexico), although recent scientific studies have revealed that the first cacao trees grew naturally in the shade of the tropical forests of the Amazon much earlier, a good 5300 years ago. It is highly probable that these spread to Central America as a result of indigenous explorations of new territories. We therefore know today that cacao has its roots in the Amazon, but it is thanks to the Olmecs, later followed by the Mayas, that it reached its highest expression both as a crop and as a cultural element. Later, the Mexica were able to fully utilise all the benefits of this sacred tree. Thus, it was in Mesoamerica that cacao became a product of great social, medicinal and economic importance, helping to increase communication routes and trade networks, as an integral part of the territorial and political expansion of the ruling populations. [L.H.–R.V.]

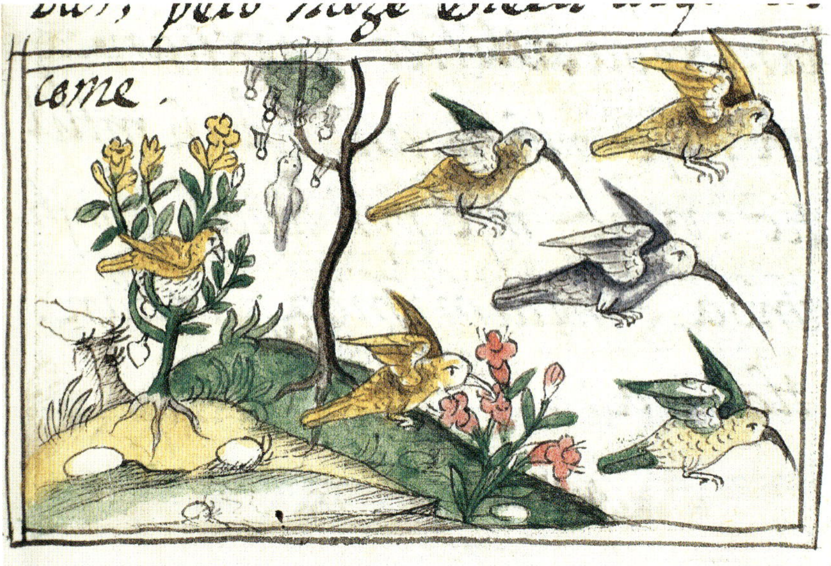

HUMMINGBIRD

Florence, Biblioteca Medicea Laurenziana, Mediceo Palatino 220

16th century (1577)
fol. 48v, detail

Mesoamerican cultures had a deep connection with nature and birds occupied an important role, symbolising concepts, taking on special meanings and religious values. They formed an essential part of their cosmogony, as they were considered the representation of certain deities. [R.V.]

Olmec Art, *Hummingbird*

Mexico, *c.* 200 BC
Olmec blue jade, 2.2 × 4.2 × 13.2 cm
Florence, Gallerie degli Uffizi, Palazzo Pitti,
Tesoro dei Granduchi, inv. Gemme 1911, no. 846

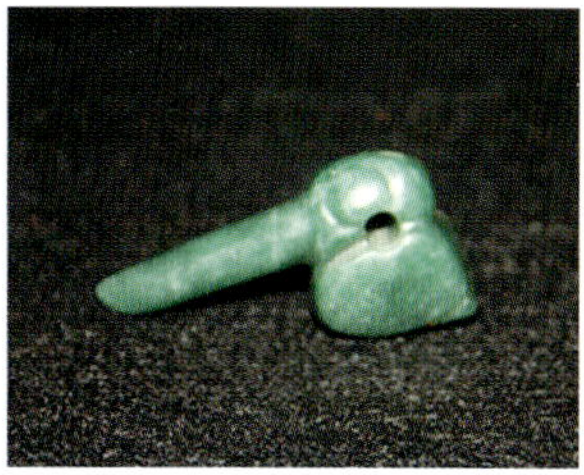

This is one of the very first indigenous objects to become part of the Medici Collection. It was identified in the inventories of the Medici 'Guardaroba' or collection in 1553 by historians Detlef Heikamp and Lia Markey. It is made of jade, a specific colour known as 'Olmec blue', and has the glassy appearance characteristic of this mineral. In addition to its long beak, one can identify the wings on the sides of the figure and above them, the eyes, drawn in a pair of circular shapes. The piece has a transverse hole on the slit dividing the bird's head from the rest of the body, which is smaller in proportion. Visible behind and below the figure is a motif of two crossed bands, known as the St Andrew's cross, a recurring motif on greenstone axes from the Mesoamerican Olmec region. It is likely that the hummingbird in the Medici Collection is an Olmec artefact from an offering of the Huey Teocalli (the Great Pyramid, or Templo Mayor of Tenochtitlán) recovered in the 16th century.

The spirituality of the Mesoamerican peoples was expressed through the production of delicate jewellery, weapons and precious objects. In addition to obsidian, gold and the feathers of particular birds, the materials used included jade as a representative element of the tireless aesthetic quest to celebrate the beauty of creation through surprising light effects that marked Mesoamerican art for millennia. The expression 'jade' refers to minerals such as nephrite, a silicate of calcium and magnesium, and jadeite, a silicate of sodium and aluminium. This mineral – which the Spaniards nicknamed 'piedra de ijada' ('stone of the hip'), due to its alleged benefits for kidneys and loins – is already archaeologically confirmed before 1400 BC, when it was used to make sharp axes for rituals. However, it was with the Olmec civilisation, which flourished from the 18th century BC onwards, that jadeite was heavily exploited, thanks to the lasting effect of their cultural action in the coastal area of the Gulf of Mexico, in the states of Veracruz and Tabasco, radiating even towards north-central Mexico and southwards in the present-day states of Oaxaca and Guerrero as far as Costa Rica. The crafting of minerals spread far and wide, eventually being particularly indicative of Mayan art as well. The artefact on display represents an ideal sample of the sophistication achieved by the Olmecs at the time of their slow and glorious political decline, which lasted until the 3rd century AD.

Found throughout the rainforests of the Mesoamerican tropics, where it originated and thrives in a variety of shapes and colours, this fascinating bird is able to adapt to a wide range of habitats. Geographically, today it can be found from Alaska to Patagonia, but its greatest diversity is concentrated along the Equator. The hummingbird family is a monophyletic group of around 340 species, characterised by variations in shape, size and behaviour. On average, it weighs only 4 grams and the length of its characteristic beak can vary from 10 to 120 mm. This tiny bird, closely tied to the figure of the Sun and its cycle of life and death, stimulated the imagination of the peoples of Mesoamerica, who were undoubtedly intrigued by the metallic reflections of its plumage and by certain behaviours that seem to break the rules that applied to all other creatures. One fact, more unique than rare, is that the hummingbird is able to hover in mid-air and fly backwards, abilities not shared with any other bird. Its plumage, capable of capturing sunlight and enhancing its beauty, seems to naturally mimic the beautiful green colour of jade, along with all shades of turquoise. This characteristic has been scientifically explained as the result of a curious physical phenomenon occurring in the 'barbules', special clusters of melanin separated by microscopic layers of air and located in the feathers: the light passing through these layers is refracted, giving the plumage a sometimes metallic and sometimes iridescent appearance and that can cover the entire spectrum of the visible.

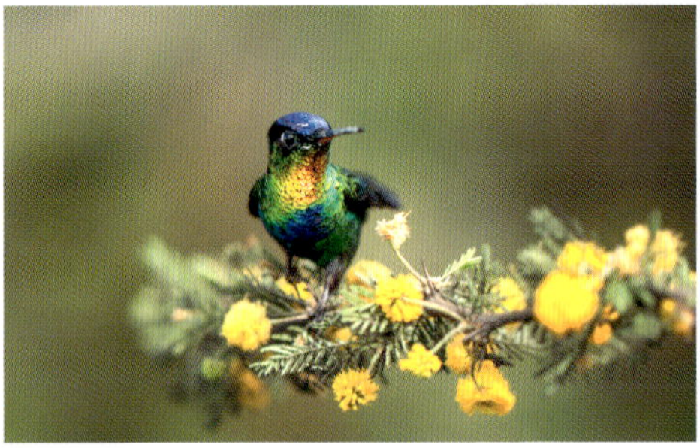

In Mesoamerican tradition it was the creature chosen to transport the souls of those who died in battle. It collected the thoughts and love of the souls of the interworld to carry them to their loved ones in the earthly realm. The Mexica associated it with Huitzilopochtli, god of the Sun and War, creator of the interworld, whose name literally means 'hummingbird from the south', one of the most important figures in cosmogony along with his brother, the sky and wind god Quetzalcóatl, who in turn was associated with another wonderful Mesoamerican bird: the quetzal. [L.H.–R.V.]

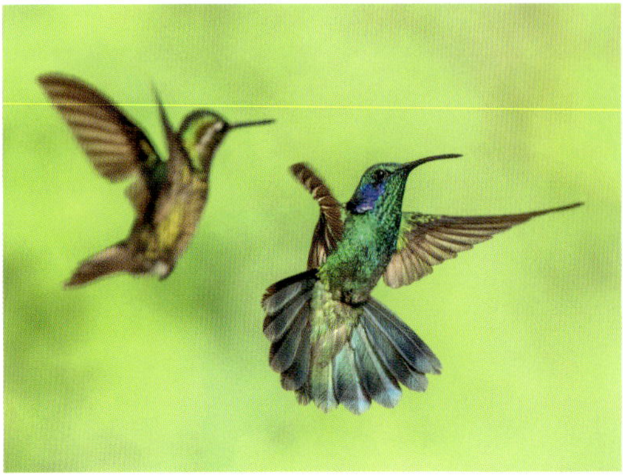

Hummingbirds.

QUETZAL

The quetzal is considered to be one of the most beautiful birds in the world; its name comes from the Nahuatl *quetzalli*, which translates as 'splendid shining feather'. It lives in the tropical forests of the Mesoamerican region, which is perhaps why it is associated with cocoa in numerous iconographic representations, because it lives in the lands where that plant grows. Its body has a kaleidoscopic iridescence that varies with the light, going from emerald green to gold and even blue, in sharp contrast to its red belly. It has a short, rounded crest on its head and a small yellow beak. Its tail, seen from above, is almost entirely white, while the upper part is covered with long green feathers that can reach 90 centimetres, when its body measures only about 35 centimetres. The quetzal was closely associated with Mesoamerican cultures as a symbol of abundance, fertility and life. *Penachos*, banners and countless garments of the Maya and Aztec cultures were made from its beautiful feathers, which were of extraordinary value in the trade of these peoples. Quetzals were not sacrificed to obtain their feathers; in fact, anyone who killed them risked the death penalty. So they were captured alive, stripped of their long tail feathers and then released back into the wild. This gave them the opportunity to grow them back during the next moult. It was therefore the first animal in history to be officially protected. Their flights of courtship are some of the most extraordinary spectacles in nature. Between February and March, during the mating season, the males capture the attention of the females through a sophisticated aerial ballet. Thanks to the sinuous movement of their long tails, from a distance it seems as though a feathered snake were flying through the air. [R.V.]

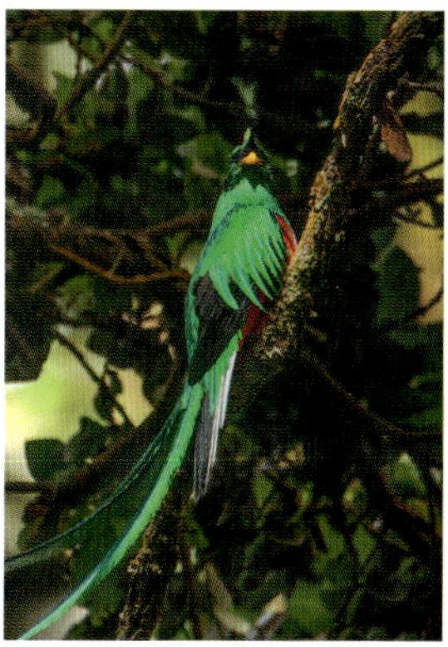

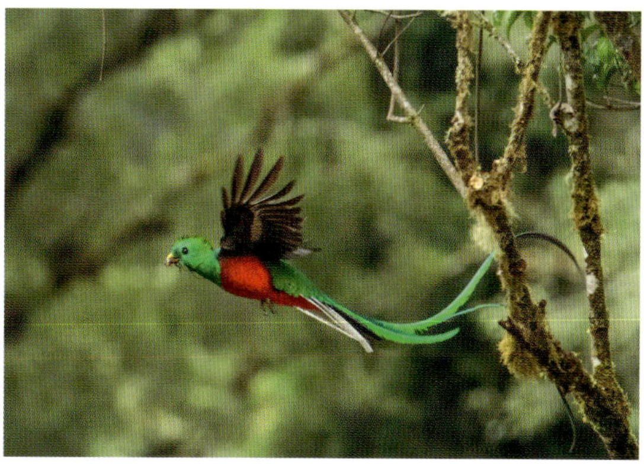

Two quetzals.

QUETZALCÓATL

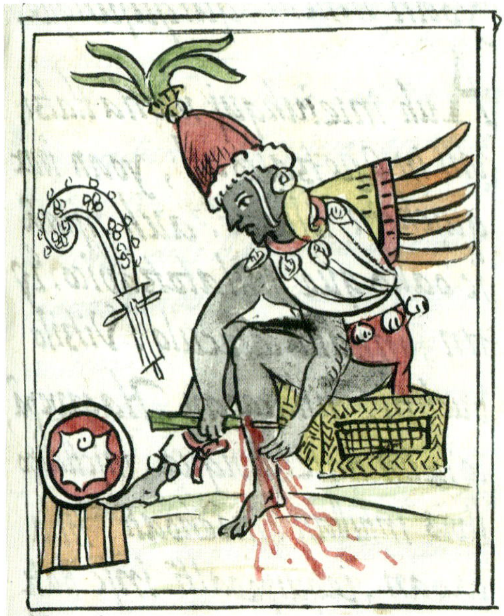

Florence, Biblioteca Medicea Laurenziana, Mediceo Palatino 218

Mexico, 16th century (1577)
fol. 36v, detail

The name Quetzalcóatl means 'Serpiente con plumas de quetzal' in Nahuatl. He was the most glorified among the Mesoamerican gods, depicted as a serpent with its head and body covered with quetzal feathers. Complex symbolism revolves around the figure of Quetzalcóatl since, in addition to being a classical deity in the Mesoamerican cosmic vision, he was a creator hero in pre-Hispanic cosmogonic mythology and was associated with figures straddling the historical and the mythological, such as Ce Acatl Topiltzin, Kukulkan, Nacxit and Q'uq'umatz. Since the Spanish conquest, the image of Quetzalcóatl – a god, myth and man who rejected human sacrifice and embodied virtues such as chastity, penitence, honesty, temperance and religiosity – has been linked with a number of diverse figures in modern history over the centuries: a Viking warrior, St Thomas, Jesus Christ and even an 'extraterrestrial' being.

The Mexica culture associates cacao with Quetzalcóatl. He was said to be Venus, the morning star, as well as god of the sky and god of the winds, and he stole from his brothers a plant that they valued for the precious drink extracted from its fruit. He descended from the sky to earth to give it to good men, in the resplendent city of Tula. Quetzalcóatl planted the cacao tree and afterwards asked Tlaloc, the god of rain, to water it so that the plant could be well-nourished and grow vigorously, and then he asked Xochiquetzal, the magnificent goddess of beauty, to bedeck it with beautiful flowers. He built his dwelling in the middle of a beautiful garden and taught the Toltec people, the inhabitants of Tula, about agriculture, astronomy, the arts and medicine. When the cacao tree bore its first fruit, he taught the women how to roast it, grind it, mix it with water in *jicaras* and make it foam, so as to obtain a divine nectar: *cacahuatl*. The gods, envious of the Toltecs' devotion to Quetzalcóatl, plotted to trick and dishonour him, using the cunning of the necromancer Tezcatlipoca. The latter, disguised as an old man, managed to persuade Quetzalcóatl to drink a mysterious potion. After drinking it, he became inebriated and lost control of his senses. The next day, disgraced and shamed by his behaviour, he decided to leave Tula. Before leaving, he set fire to his house and turned the cacao into a tree bereft of flowers and fruit. Then he ordered the wonderful birds with shining feathers to leave the area. Distraught, but vowing to return for revenge, he headed for the East by a sea route.

In the future, on the other side of the world, way beyond that East to which Quetzalcóatl had fled, following the unheard-of proposal of the Florentine Paolo dal Pozzo Toscanelli (1397–1482), mathematician, astronomer, geographer and cartographer in the service of Cosimo de' Medici the Elder (1389–1464), what Tzvetan Todorov calls 'the most astonishing encounter of our history' in his book *La conquête de l'Amérique. La question de l'autre* (Paris, 1982) was to take place. His suggestion to sail westwards to reach the Orient was to inspire Columbus' expedition.

With catastrophic results.

Mixteca-Puebla art, *Mask*

Mexico, 15th–16th century
Wood, plant resin, turquoises, spondylus, mother of pearl, 24 × 15.8 cm
Rome, Museo delle Civiltà, inv. 4213

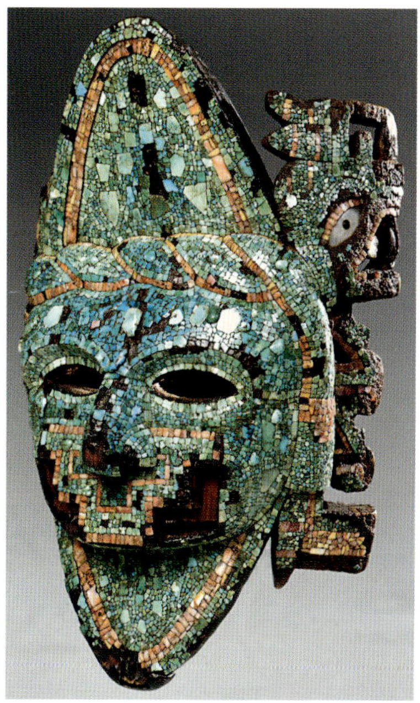

The mask is decorated with tiles of turquoise, jadeite, mother-of-pearl, coral, shells and semiprecious stones. It is one of the most beautiful artefacts of the Aztec culture, but is in the traditional style of the Mixteca-Puebla, which excelled in creating various types of mosaic objects. Of the objects produced with this technique, of which there are no more than twenty-five in the world, it is not only one of the most valuable, but also the best preserved. The mask represents a face, with hollow eyes, nose and mouth, sandwiched between two triangular mandibles of a snake, whose gums are portrayed with strips of red shells. The face is adorned with a *nariguera* (nose pendant), while the forehead has a decoration consisting of intertwined snakes, whose broad heads are placed on both sides of the mask: the one protruding on the left side has feather-like appendages. The identity of the character represented in the mask is uncertain, believed to be either Quetzalcóatl, the Plumed Serpent, a civilising deity, or Chalchiuhtlicue, the goddess of water and fertility, or Xiuhcóatl, a zoomorphic disguise of the fire deity Xiuhtecuhtli. The object may have been made between 1461 and 1521, perhaps in the Valley of Mexico or the coast of the Gulf of Mexico, where, at the height of the Aztec empire's expansion, the influence of the Mixteca-Puebla stylistic tradition had spread. The mask may have come to Europe with the objects donated to Cortés by Moctezuma II, the Aztec ruler, and sent by them to King Charles V (1500–58) in 1519–20, or with those collected by Juan de Grijalva during his journey from Yucatán to northern Veracruz in 1518 and presented to the king in Barcelona in 1519.

It is essential to include a scientific insight into the material used in the work: turquoise. The most iconic forms of Mesoamerican art are made from this precious material. An international group of geochemists has determined that the real place of origin of turquoise is Mesoamerica, according to a study published in June 2018 in the specialist journal *Science Advances*. This study contradicts the hitherto accepted version that claimed that ancient Mesoamerican civilisations acquired this stone by importing it from the southwestern United States of America. Scientists conducted a comprehensive lead and strontium isotope analysis on mosaic fragments with pieces of turquoise on valuable objects from the Mexica and Mixteca cultures. The study revealed that these pieces had isotopic signatures consistent with Mesoamerican geology, and not that of the southwestern United States. [R.V.]

TULA

The identification of Quetzalcóatl with the Toltec king Ce Ácatl Topiltzin, who actually lived in the 10th century, underpins several legends about the god. Tula, capital of the Toltec state, flourished between the 10th and 12th centuries in the western part of the present Mexican state of Hidalgo. The myth points to Quetzalcóatl as the founder of Tula, which at its heights had as many as 35,000 inhabitants and a reach that went as far as Nicaragua, El Salvador and the Yucatán peninsula. At its peak, Tula boasted a series of impressive temples, which, according to legends and testimonies, were gaudily decorated. One temple was said to be entirely covered in turquoise, one in shells, one in gold and one even covered in precious quetzal feathers.

The Atlantean columns of Tula are imposing basalt stone sculptures over 4.5 metres high, representing Toltec warriors. Decorated with symbolic elements, including a breastplate in the shape of a butterfly, they originally supported the roof of the temple dedicated to Tlahuizcalpantecuhtli, who was none other than a depiction of Venus, the morning star, and one of the aspects of Quetzalcóatl. As we shall see, the credit the Mexica gave to his myth was responsible for the reception they gave to the Spaniards. [L.H.]

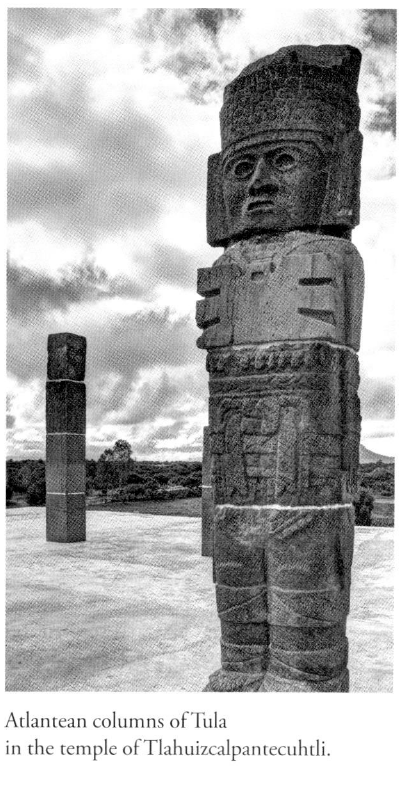

Atlantean columns of Tula
in the temple of Tlahuizcalpantecuhtli.

2. DIVINATION

Book IV, Astrology and Divination
'de la astrología judiciaria o arte adivinatoria indiana'

Book V, Superstition
'que trata de los agüeros y prenósticos que estos naturales tomaban de algunas aves, animales y sabandixas para adivinar las cosas futuras'

Book VI, Language and Moral doctrine
'de la retórica y filosofía moral y teología de la gente mexicana, donde hay cosas muy curiosas tocantes a los primores de su lengua y cosas muy delicadas tocantes a las virtudes morales'

Book VII, Sun, Moon, Stars
'que trata de la astrología y filosofía natural que alcanzaron estos naturales de esta Nueva España'

Man has always had a natural need to explore and discover. While today we are able to see the earth from space, man once used to look to the sky and its stars to orient himself in the world – the theatre in which history unfolds. One of the tools Geography uses to define, navigate and learn about the world is Cartography. A number of maps of great historical value date back to the Middle Ages and the modern age, including the map reconstructed on Ptolemy's *Geographia*, numerous portolans and the so-called 'world map of Fra Mauro', the richest map in terms of style, content and geographical perspective. [R.V.]

Fra Mauro's world map

Venice, 1450
Paint and gold on parchment glued to wood, 230 × 230 cm
Biblioteca Nazionale Marciana, Venice

Fra Mauro's world map, a multi-media exploration by the Museo Galileo in Florence
by Filippo Camerota and Angelo Cattaneo

The mid-15th century was characterised by competition between European powers, particularly Spain and Portugal, for control of maritime trade routes. The quest for new routes to the East led to a significant revision of the maps of the time.

The world map by Fra Mauro (1400–*c.* 1460?) in the Biblioteca Nazionale Marciana in Venice is one of the most fascinating cartographic products of the 15th century. Made in the Camaldolese monastery of San Michele in Isola, Venice, around 1450, it represents a bridge between mediaeval geographical knowledge and the commercial enterprises that would lead to the discovery of the New World a few decades later. In 1942, the Italian government wished to present it at the Universal Exhibition in Rome as an eloquent example of Italy's contribution to the great geographical explorations, and for that occasion had a facsimile reproduction made. Today it is on display in the collections of the Museo Galileo in Florence. The facsimile was made in Florence by the Laboratorio Alinari. Using the collotype technique, they succeeded in perfectly reproducing the rich chromatic quality of the great painting on parchment, where the geographical representation of seas and continents is enriched by a large number of figurative elements, inscriptions and cosmographic schemes enclosed in a splendid gilded frame.

The beauty of the work, with which visitors are unfailingly captivated, however, is countered by the obvious fact that the modern observer finds it difficult to read, completely disorientated by the upside-down representation of the world and the almost three thousand inscriptions in Venetian vernacular that cover the entire geographical representation in the form of cartouches. To make the work easier to understand – in agreement with

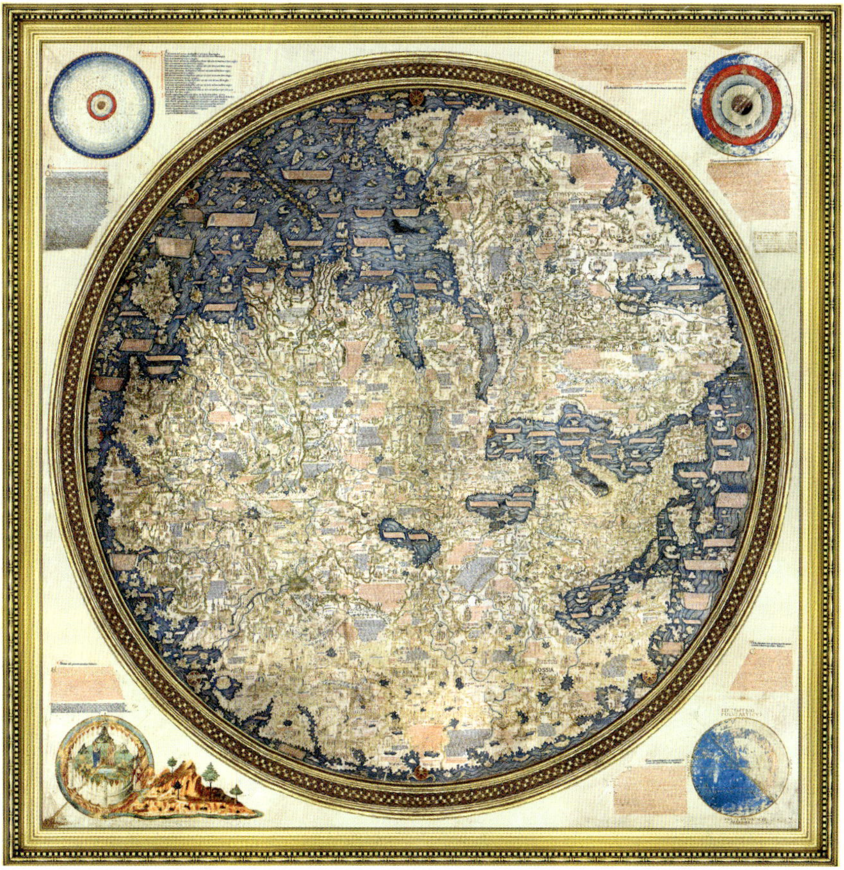

the Biblioteca Nazionale Marciana, which recently restored the beautiful world map – the Museo Galileo developed a website that uses multimedia to explore it, modelled on the similar site published in 2016 on Martin Waldseemüller's (*c.* 1470–1521) large world map.

Although produced in different times and places – Waldseemüller's planisphere was printed in Lorraine, in Saint-Dié-des-Vosges – these two monumental maps of the world may have a connection to the work of a famous German cosmographer who was active in Florence at the time of Lorenzo the Magnificent (1449–92), Heinrich Hammer (1440–96). He was author of a planisphere that dates to slightly before the discovery of the New World, but is crucial to the drafting of Waldseemüller's map. Hammer's planisphere was drawn up around 1490, when there was a manuscript copy of Fra Mauro's world map in the Palazzo dei Medici, presumably made a few years earlier at the behest of Lorenzo the Magnificent.

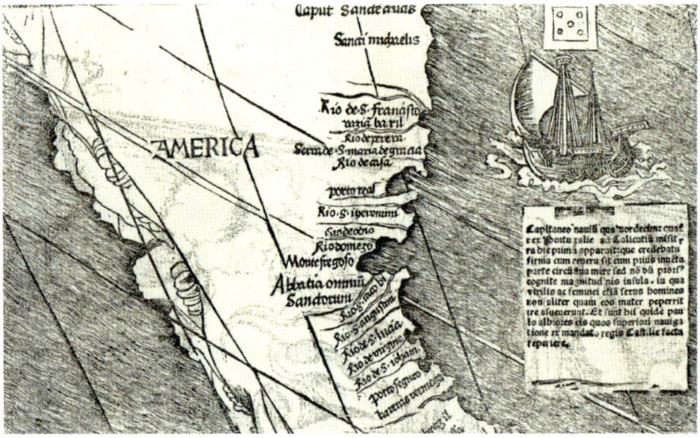

Martin Waldseemüller, *Universalis Cosmographia secundum Ptholomaei traditionem et Americi Vespucii aliorumque lustrationes*, detail, Saint-Dié-des-Vosges, 1507, Washington, D.C., Library of Congress, Geography and Map Division.

Hammer was one of the first cosmographers to depict the complete coastal outline of Africa, following the first famous circumnavigation by the Portuguese Bartolomeu Dias (1450–1500) in 1488, extending the ecumenical space beyond the geographical limits indicated by Ptolemy, as far as the still unknown island of 'Cipangu', or the Japan described by Marco Polo (1254–1324). About half a century earlier, Fra Mauro himself had imagined and prefigured this extraordinary cosmographic scenario in his world map.

Fra Mauro's world map is a representation of the *imago mundi* of the mid-15th century, a figurative summation of the geographical novelties and 'literary' discoveries available at the time. Painted and historiated in bright colours, it is inscribed within a circle approximately two metres in diameter. The geographical representation is enriched with over three thousand cartouches, many place names and hundreds of images of cities, temples, roads, ships and a beautiful earthly paradise illuminated by Leonardo Bellini (1423–?).

The earthly paradise is a legendary place that mediaeval geography placed in the far East, on a distant island located in the direction of the sunrise; Fra Mauro places it in cosmographic space, depicting it outside the ecumene, along with celestial diagrams illustrating the spheres of the planets, the elements, tidal theory and other elements of natural philosophy. Taken as a whole, the historiated map, with diagrams of the sky and commentary on it in cartouches summarising famous mediaeval philosophical texts, appears as a kind of figurative treatise on cosmography. Fra Mauro outlines the image of the world just before the Portuguese and Spanish navigations, integrating the *Geographia* of Ptolemy (*c.* 100–*c.* 175 AD) with the travel accounts of Marco Polo and Niccolò di Giovanni Conti (1395–1469). The ancient ecumene expands eastwards as far as Japan and southwards as far as the southernmost latitudes of Africa, which, although not completed, clearly hints at the possibility of its circumnavigation.

Marco Polo's travel accounts, collected in the famous *Milione* [*The Travels of Marco Polo*], were written after his return to Venice, probably during his imprisonment in Genoa in 1298, in collaboration with the poet Rustichello da Pisa (second half of the 13th century).

Made public around 1300, the manuscripts of the *Milione* were in a variety of languages and were broadly circulated, disseminating the characteristics of a true economic and political geography of Asia described as a dense and intricate network of land, river and sea trade routes. Following this, the Mediterranean and Christian Europe were connected to the deepest and most distant Asia.

One of the Venetian merchants who contributed to the knowledge of the geography of Asia is Niccolò di Giovanni Conti. In 1414, he embarked on a long journey to India, Indochina and Borneo. His mastery of Arabic and Persian allowed him to acquire an in-depth knowledge of the territories less explored by Marco Polo. He returned to Venice twenty-five years later, and Pope Eugenius IV (1383–1447) obliged him to give an account of his travels as part of the penance imposed on him for abjuring the Christian religion. The account was dictated to Pope Francis' secretary, Poggio Bracciolini (1380–1459), who made it into *Liber IV* of *De varietate fortune*, which was made public in 1448 and circulated through numerous manuscript copies and some vernacularisations in Tuscan and Venetian. The wealth of information about India and the many islands of the Indian Ocean constituted, together with the tales of Marco Polo, a valuable source of information for Fra Mauro's world map. In addition to having imagined and outlined the concrete possibility of joining the Mediterranean to the Indian Ocean, circumnavigating Africa in one immense route, Fra Mauro drew an extraordinary and very detailed representation of the inner regions of Africa. The representation of Ethiopia, in particular, remained unsurpassed for hundreds of years. Fra Mauro made use of maps he received from some Ethiopian monks who had come to Venice – probably on the occasion of the Council of Florence (1439) – and collected their accounts and testimonies. Based on this information, he depicted the Ethiopian kingdoms far beyond the latitudes that Ptolemy believed to be habitable.

After his death, Fra Mauro was quickly forgotten and for about three centuries it was believed that the world map of the Camaldolese monastery of San Michele in Isola was drawn by an unknown lay brother who had copied a map of the world drawn up by Marco Polo during his long journey to the lands of the Orient. Fra Mauro's name was only rediscovered in the mid-18th century in the context of historical studies on the Camaldolese Congregation of the Order of St Benedict. It has since been subjected to careful critical study, the results of which are now on the website designed and built by the Museo Galileo in Florence. The website makes it possible to explore the extremely rich textual and graphic information of the world map in detail, displaying the approximately 3,000 cartouches describing kingdoms and provinces, selecting seas and rivers, cities and monuments, roads and land and sea trade routes. Most of the places described by Marco Polo have been identified and geolocated on the Google map so that the route taken by Marco, his father and uncle can be reconstructed. Niccolò de' Conti's travel route was visualised in the same way, as were the routes of the Portuguese, who shortly after the drafting of the world map were able to reach the Indies by sea, rounding the Cape of Good Hope and connecting to a route, indicated by Fra Mauro, that from that southernmost point, the Chinese junks had already traced as far as the coasts of India and Catai. The Turkish blockade of the Mediterranean, imposed with the fall of Constantinople in 1458, had in fact obliged the maritime powers to seek an alternative route to the East, and Fra Mauro's map was certainly the most suitable cartographic document for developing such a commercial enterprise. Not surprisingly, the king of Portugal requested a copy, now lost, which was sent to Lisbon before 1459.

From an iconographic point of view, Fra Mauro's world map falls within the tradition of the Mediaeval *mappaemundi* and portolan charts. Based on a specific choice which the Venetian cosmographer accounted for in one of the cartouches, the drawing of the world does not follow Ptolemy's humanistic tradition, which established the position of places by means of the geographical coordinates of latitude and longitude. Instead, Fra Mauro inscribes the known world within a wind rose, as was customary in nautical charts, and draws regions, cities and orographical details by compiling information from portolans and travel reports of merchants and missionaries. The coordinates of the places are based on compass bearings and therefore the entire world map shows the same anti-clockwise rotation of about eleven degrees that characterises nautical charts; a rotation due to the magnetic declination of the compass needle whose direction, as is well known, differs by several degrees from that of geographical north. The compelling animations developed for the website effectively illustrate these aspects of mediaeval cartography, a subject that today is more commonly studied by specialists in the field.

Also from Fra Mauro's workshop is a beautiful nautical chart, today in the Vatican Library. What distinguishes it from the nautical charts of the time is its wealth of cosmographic content. Not only are the cartouches similar to the Venetian *mappamondo*, but also the drawing of the Mediterranean coastline, which is completely identical in size as well. The two maps are perfectly superimposable, to the extent that the scale of distances in nautical miles drawn on the nautical chart, according to the custom of the time, is legitimately transferable to the world map, allowing us to measure the size of the then known world. By relating the lands depicted on the world map to their actual geographical extent, from the Canary Islands to Japan, the world depicted by Fra Mauro should today extend for approximately 160 degrees of longitude.

However, like most Renaissance cosmographers, Fra Mauro believed Eurasia to be much more extensive; the cosmographer therefore compressed Ptolemy's eastern regions to include 'Mangi' and 'Cathaio', or China and Japan, territories unknown to the Alexandrian scientist and described by modern scientists. In the circular shape of the world map, in addition to the 180 degrees of longitude of the Ptolemaic ecumene, the world he depicts turns out to have a diameter of about 11,500 km, or a circumference of about 36,000 km, a figure that lies somewhere between the 33,000 km of Ptolemy (2nd century) and the 40,000 of Eratosthenes (3rd century BC).

Attributed to Johannes Vingboons and workshop, after Hessel Gerritsz, 1630
Map of the Santo Francisco port on the north coast of Hispaniola

Amsterdam, *c.* 1630–60
Line-and-wash drawing on canvas-backed paper, 61.9 × 47 cm
Florence, Biblioteca Medicea Laurenziana, Carte di Castello, 42

The Genoese explorer and navigator Cristoforo Colombo or Christopher Columbus (1451–1506), in an attempt to reach the Orient by travelling west, sighted an unknown land on the fateful 12 October 1492 that would radically change the geographical knowledge of the world and the course of history. On that day he landed on the island of Guanahaní, which he christened San Salvador, in the Antilles in the Caribbean Sea. He later arrived on the island he christened 'La Española' (Hispaniola, in today's territories of Santo Domingo and Haiti). 'La Española' was the first territory conquered and colonised by Spain in the Americas. There, the Crown experimented with and tested colonial projects in the New World and established the first political, legal, economic and cultural institutions.

Columbus was the first European to come into contact with cocoa during his fourth voyage to what he mistook for the East Indies. His son Fernando recounts this in a letter published in Italian (the original in Spanish has been lost) in Venice, in 1571: on 15 August 1502 on the island of Guanaja (Bay Islands, Honduras), they met indigenous Maya Chontales, originally from the territory that is now Tabasco in Mexico, on a boat full of trading goods, including 'una buena cantidad de esas almendras' ('a good amount of those almonds'). They were cocoa beans.

Since he had no interpreter, the admiral did not understand their value and did not know that these almonds were used to make chocolate. [R.V.]

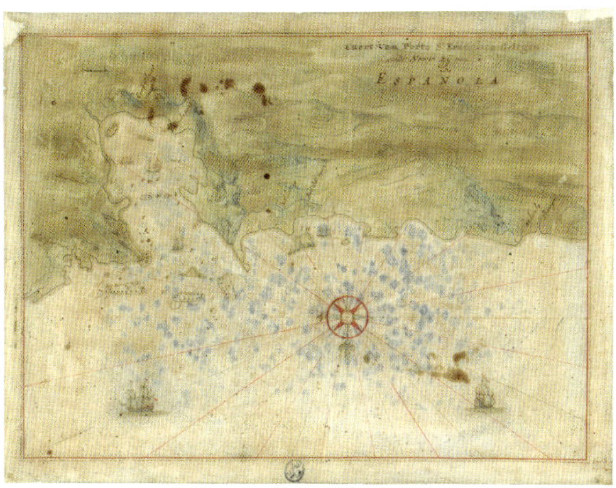

Martin Waldseemüller, *Universalis Cosmographia secundum Ptholomaei traditionem et Americi Vespucii aliorumque lustrationes*

Saint-Dié-des-Vosges, 1507
Map on 12 sheets of paper, created from an original wood engraving, 129 × 232 cm
Washington, D.C., Library of Congress, Geography and Map Division

Martin Walseemüller's planisphere, a multimedia exploration by Museo Galileo, Florence
by Filippo Camerota

The planisphere by the German cartographer Martin Waldseemüller (*c.* 1470–1521) is the result of an ambitious project developed in the early 16th century in the small Lorraine village of Saint-Dié-des-Vosges near Strasbourg. Supported by the Duke of Lorraine, René II (1451–1508) – who in 1490 had founded a centre for humanistic studies on Ptolemy's *Geographia* – the project involved documenting and updating geographical knowledge derived from recent Spanish and Portuguese overseas navigation. At the origin of this idea was undoubtedly the interest aroused by Amerigo Vespucci's travel accounts: the letter to Lorenzo di Pierfrancesco de' Medici (1463–1503) known as *Mundus Novus*, published in 1504, and the letter to Pier Soderini (1450–1522), which in the same year illustrated the results of the four voyages made by the Florentine navigator between 1497 and 1504. Those letters revealed the existence of a land that extended 'beyond the stars' of the northern hemisphere, beyond the Tropic of Capricorn, for at least another 20 degrees of latitude, prompting the urgency of an overall redrawing of the world map. The project almost certainly took shape in 1505, when René II received from Portugal a manuscript copy of the letter to Soderini, translated into French and dedicated to him. The humanist Matthias Ringmann (1482–1511), who had published a Latin edition of the *Mundus Novus* that same year, was then called upon to take charge of the project in collaboration with the cartographer whom the humanists of Saint-Dié identified as the most competent on the subject, Martin Waldseemüller.

Out of that collaboration came one of the most interesting published works of the time: a pamphlet introducing cosmography (*Cosmographiae Introductio*), almost certainly written, at least in large part, by Matthias Ringmann; a Latin edition of the letter to Soderini, translated from the French version dedicated to René II and published at the end of the cosmographic pamphlet; and two world maps drawn by Waldseemüller reproducing in flat and solid form almost the entire globe, with the exception of the southern cap. These were, respectively, the great 'Universal cosmography according to the teachings of Ptolemy and the voyages of Amerigo Vespucci and other explorers', and a small globe that for the first time was represented in the form of twelve spindles. The two maps represented an

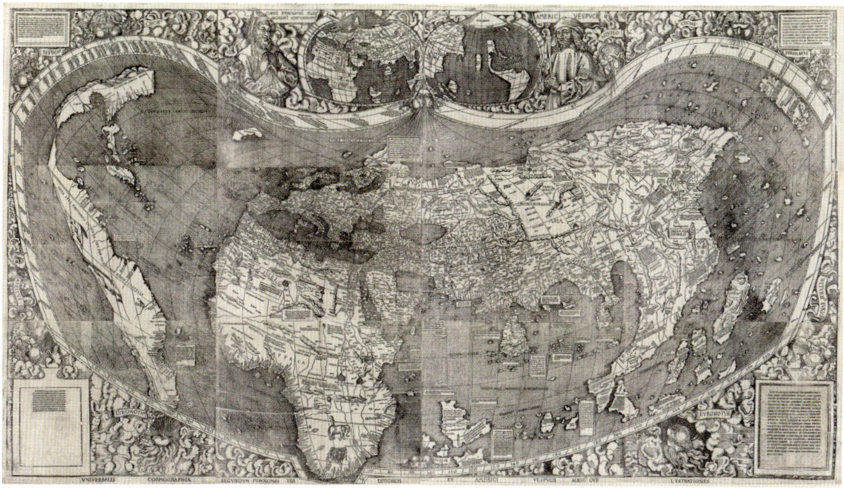

extended world that included, beyond the limits of the Ptolemaic ecumene, the lands described by Marco Polo and, above all, those recently discovered by Christopher Columbus and Amerigo Vespucci. And it was precisely in honour of Vespucci, the first man able to recognise the recently explored coastlines as part of a new continent, that Ringmann and Waldseemüller baptised those lands with the name 'America'. For this reason, the map is known today as the 'birth certificate' of America.

By depicting the New World as a hitherto unknown separate continent, Waldseemüller's map documented Vespucci's revolutionary idea, forever changing the traditional tripartition of the world into Europe, Asia and Africa. Formed by assembling twelve sheets measuring approximately 43 × 58 cm (1.5 × 2 Lorenese feet), the *Universalis Cosmographia* measures approximately 129 cm high by 232 cm wide, making it one of the most challenging cartographic products of the modern age. The specimen in the Library of Congress is the only surviving copy of a print run that, according to the author, numbered as many as a thousand. Long considered lost, the map was found in 1901 by the Jesuit historian and cartographer Joseph Fischer (1858–1944) in the library of the castle of Prince Johannes of Waldburg-Wolfegg (1833–1906) in Baden-Württemberg. It was acquired in 2001 by the Library of Congress of the United States, and is now on display in the Great Hall of the Thomas Jefferson Building.

(by Filippo Camerota, *Una terra oltre le stelle. Il planisfero di Martin Waldseemüller, 1507*, Florence, 2016).

Attributed to Johannes Vingboons and workshop, after Hessel Gerritsz, 1630
Isla de la Vaca map and bay at la Hispaniola

Amsterdam, *c.* 1665
Line-and-wash drawing on canvas-backed paper, 61.9 × 47 cm
Florence, Biblioteca Medicea Laurenziana, Carte di Castello, 50

The map depicts the small Isla de la Vaca in the south-west of Haiti in the Caribbean Sea and the southern part of the Haitian coast currently occupied by the town of Les Cayes. In the 17th century, it was the hideout of the pirate Henry Morgan (1635–88).

At that time, all the natives were enslaved: they mainly mined gold and cultivated sugar cane, the productive activities that contributed to the wealth of the Spanish Crown. It was precisely from Isla de la Vaca, the first to be conquered, colonised and enslaved, that the greatest demand for freedom in the New World arose. In fact, the independence of the United States did not involve its considerable enslaved population. Haiti was actually the first state to gain its freedom. François Dominique Toussaint-Louverture (1743–1803), leader of Haiti's independence and anti-slavery revolution of 1791, is responsible for laying the foundations for the ultimate elimination of slavery on the American continent and, consequently, worldwide. [R.V.]

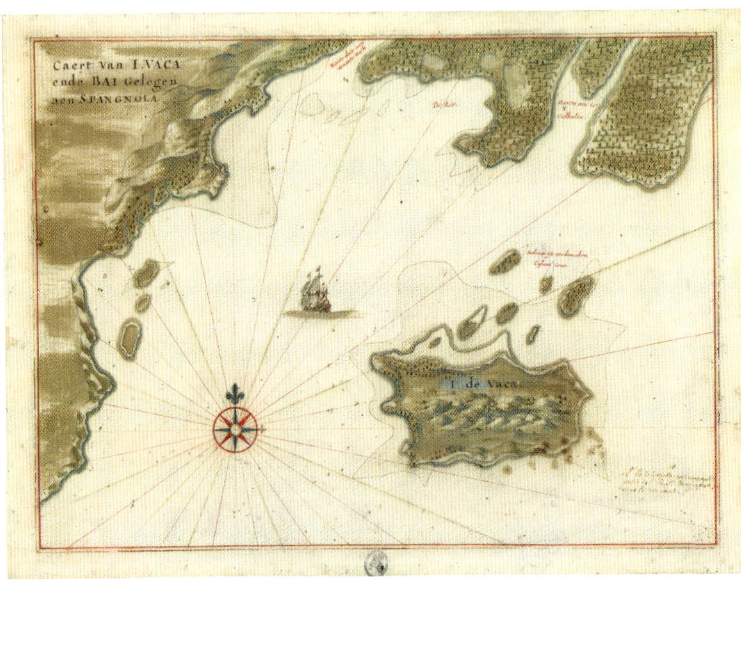

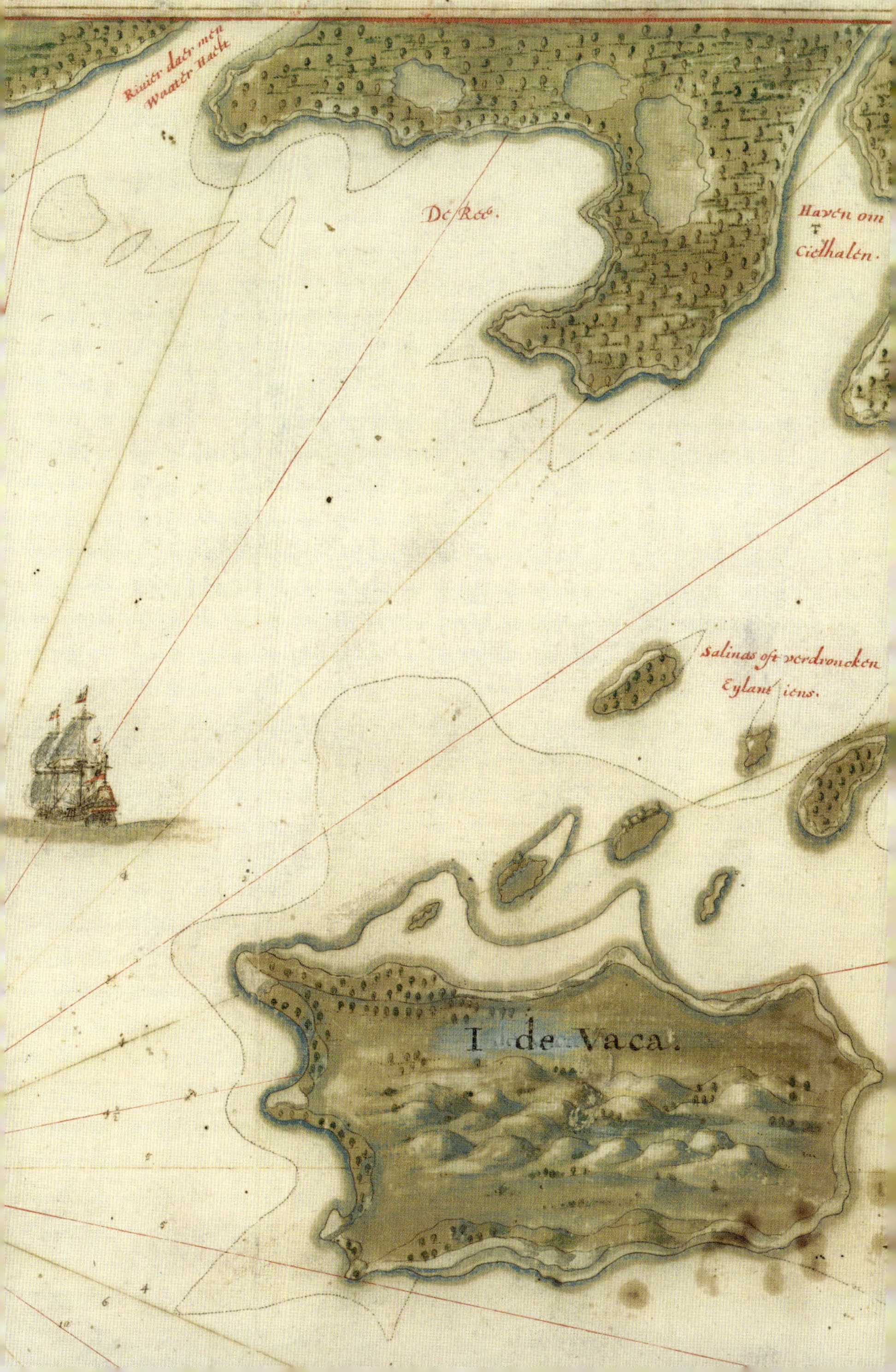

Riuier daer men
Waater haelt
De Reé.
Haven om
Cielhalen.
Salinas oft verdroncken
Eylant iens.
I de Vaca.

TENOCHTITLÁN

Aztec migration and foundation of Tenochtitlán

in Giovanni Francesco Gemelli Careri, *Giro del mondo*, 6 vols., Naples, 1699–1700, VI, plate facing p. 38

At the time of the discovery of the New World and the colonisation of Haiti, the hegemonic civilisation in Mesoamerica was the Mexica people. Their ruler was Moctezuma II, the leader of an empire of millions of subjects in the heart of the city of Tenochtitlán, with a strong warrior tradition. He was the second *tlatoani* (literally 'he who can speak', in other words, the emperor) to have this name. He was the son of Axayácatl – sixth *tlatoani* of the Aztecs – and apparently grandson of Nezahualcóyotl, poet-king of Texcoco. In 1502 he was appointed supreme ruler and, after the *xiuhuitzolli* (the crown of power and time) was placed on his head, he was told, 'You must go out and look at the stars to know the times and the signs they bring and their influences and what they threaten. And you must pay attention to the planet Venus, the morning star'.

The Mexica Empire was a political and military alliance of three city-states: Tenochtitlán, Texcoco and Tlacopan, the 'Triple Alianza'. Tenochtitlán, located on an island in Lake Texcoco, like a surprising Mesoamerican Venice, was the capital. A powerful theocratic-military government exercised hegemony over the peoples of the plateau, forcing them with weapons to pay heavy tributes. Of all the products that were recognised, gold, feathers and cocoa were the most significant; all the goods received went directly into the imperial economy for commercial purposes, creating the most impressive economic structure in all of Mesoamerica. [R.V.]

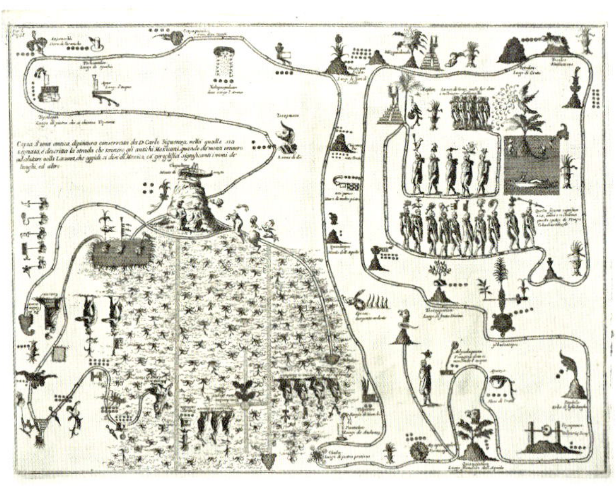

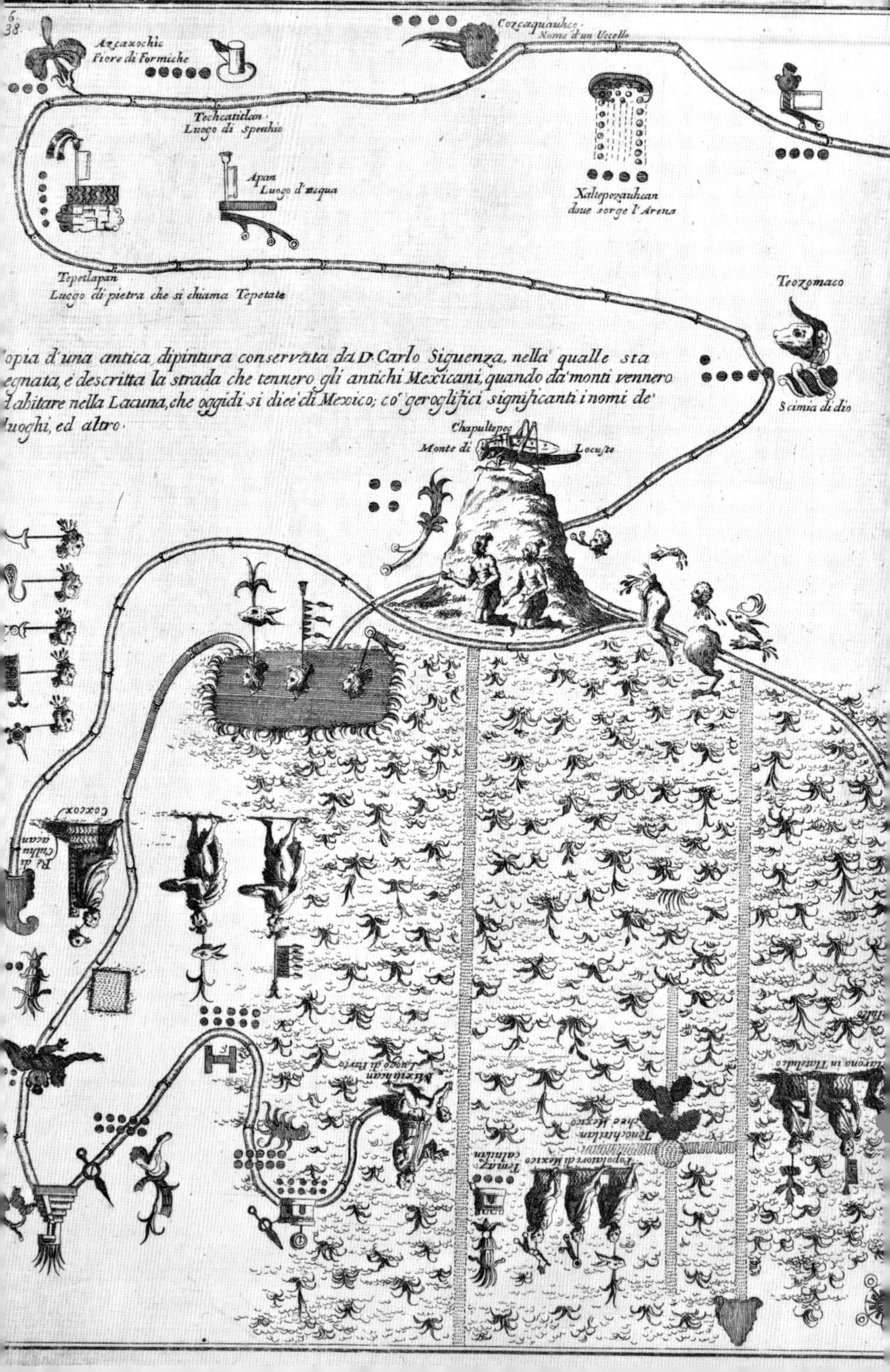
Azcaxochic
Fiore di Formiche
Cozcaquauhco
Nome d'un Uccello
Techcatitlan
Luogo di Specchio
Apan
Luogo d'acqua
Xaltepozauhcan
doue sorge l'Arena
Tepetlapan
Luogo di pietra che si chiama Tepetate
Teozomaco
Scimia di dio
opia d'una antica dipintura conservata da D. Carlo Siguenza, nella qualle sta
egnata, e descritta la strada che tennero gli antichi Mexicani, quando da' monti vennero
d abitare nella Lacuna, che oggidi si dice di Mexico; co' geroglifici significanti i nomi de'
luoghi, ed altro.
Chapultepec
Monte di Locuste

Florence, Biblioteca Medicea Laurenziana, Mediceo Palatino 219

Mexico, 16th century (1577)
fols. 252v, 262r–v, 263r, details

The civilisations of Mesoamerica believed strongly in premonitions and omens of the future. They observed the sky in a systematic way and the recording of events related to celestial bodies was carried out by priests by monitoring and predicting the phases of the Moon, eclipses and when the celestial bodies most important to them appeared and disappeared, such as Venus.

From 1508 and with the approach of the Acatl year 1 of the Mexican calendar, corresponding to 1519, Moctezuma II noticed a series of events that were interpreted as omens. Among these events were some 'signs', such as the appearance of a bright star, rising from the East and setting precisely in Mexico, falling like a shower of sparks, and the discovery of a bird that had a mirror in its head that reflected the sky and the stars. Since 1517, Moctezuma II had also taken to watching the coast, fearing that the various omens heralded the imminent end of his reign. [R.V.]

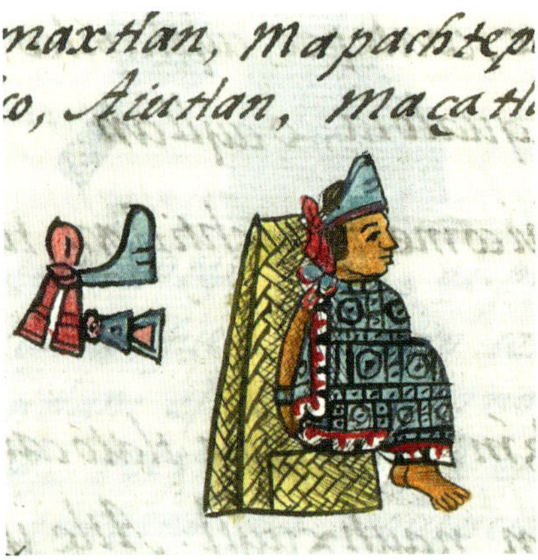

Detail of fol. 252v.

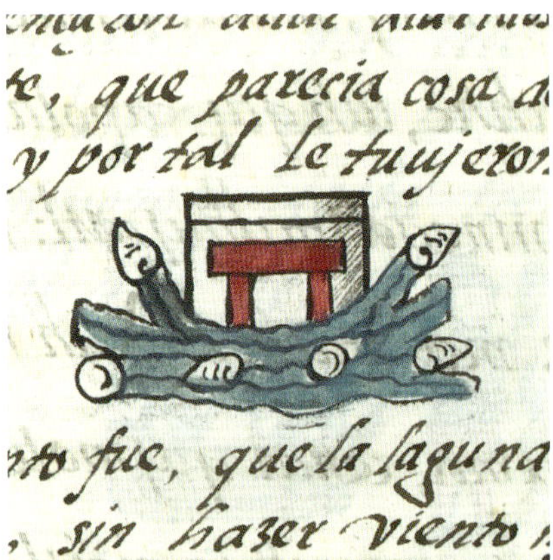

Detail of fol. 262r.

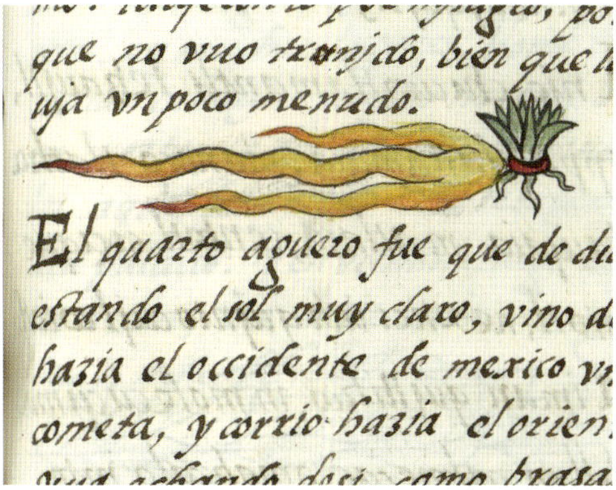

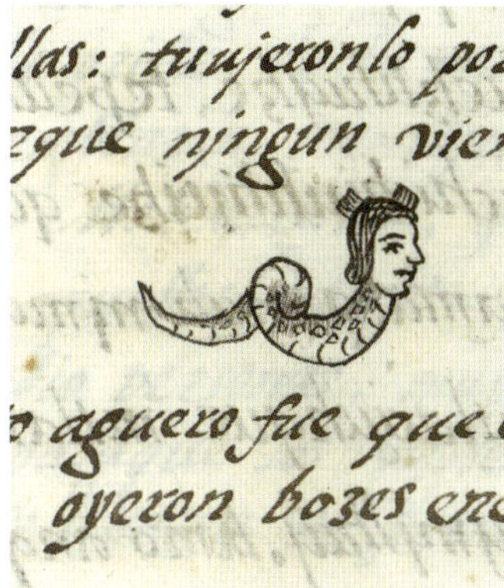

Details of fol. 262r.

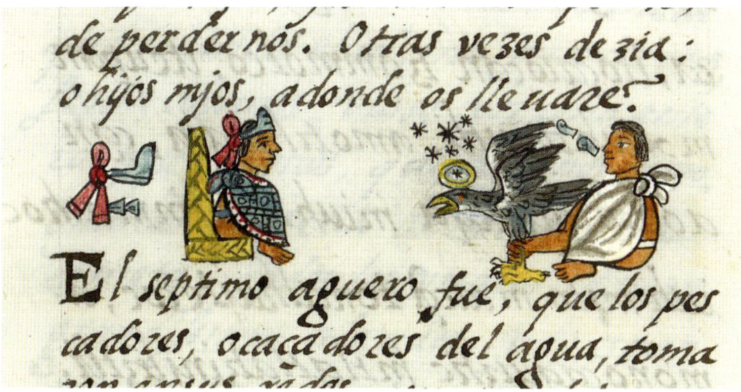

Detail of fol. 262v.

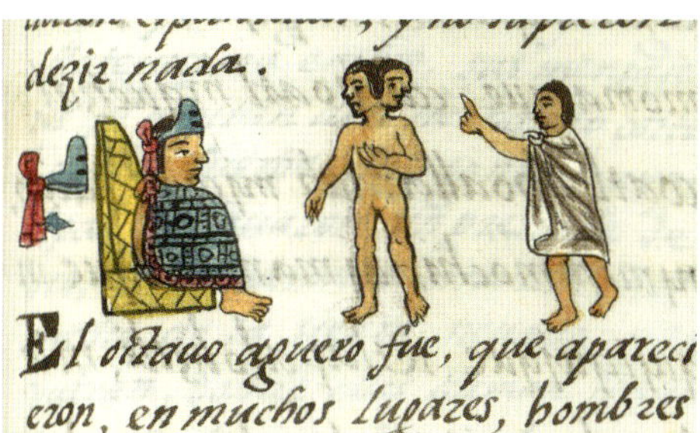

Detail of fol. 263r.

HERNÁN CORTÉS

Cristofano dell'Altissimo, ***Portrait of Hernán Cortés***

1552/68
Oil on canvas, 60 × 44 cm
Florence, Gallerie degli Uffizi, Galleria delle Statue e delle Pitture, inv. 1890, no. 53

The portrait, part of the Jovian Series in the Uffizi, depicts Hernán Cortés (1485–1547, latinized as Ferdinandus Cortesius), a Spanish conquistador who became governor of New Spain at the behest of Charles V in 1522, after having participated in the conquest of Cuba in 1511 and captained the conquest of Mexico in 1521. The figure, with a long grizzled beard and dark hair, wears a black cap and a brown jacket from which part of the light collar of his shirt, with decorative motifs, emerges. The painting is set in a natural walnut wooden frame with a carved ornamental motif of ovules, a type of moulding with a convex quarter-circle edge widely used in Greco-Roman architectural decoration, alternating with arrow-shaped darts. The internal and external edges, as well as the carvings, are gilded.

Few people, taken individually, can be said to have had as much impact on the course of historical events as Hernán Cortés. He was first and foremost a typical exponent of the minor Spanish nobility, a kind of ideal model for understanding the anthropological and cultural context underpinning the development of European imperialism between the Middle Ages and the modern age. Born in Medellín, in the province of Extremadura, to Martin Cortés de Monroy, a *hidalgo* who put his mother's surname before his own for tax reasons, Cortés came from a prestigious dynasty of men-at-arms. A second cousin of Francisco Pizarro – the equally infamous conqueror of the Inca Empire –, restless and eager to make his fortune, he embarked for Santo Domingo in 1504, the year of the death of Queen Isabella the Catholic, and then followed Diego Velázquez in the conquest of Cuba in 1511. In 1519, he arrived in Mexico to conduct his campaign of conquest. Emperor Charles V granted Hernán Cortés the Marquisate of the Valley of Oaxaca and for a time he lived as a great lord. Returning to Spain and overwhelmed by intrigues, he died in obscurity in 1547 near Seville. [L.H.–R.V.]

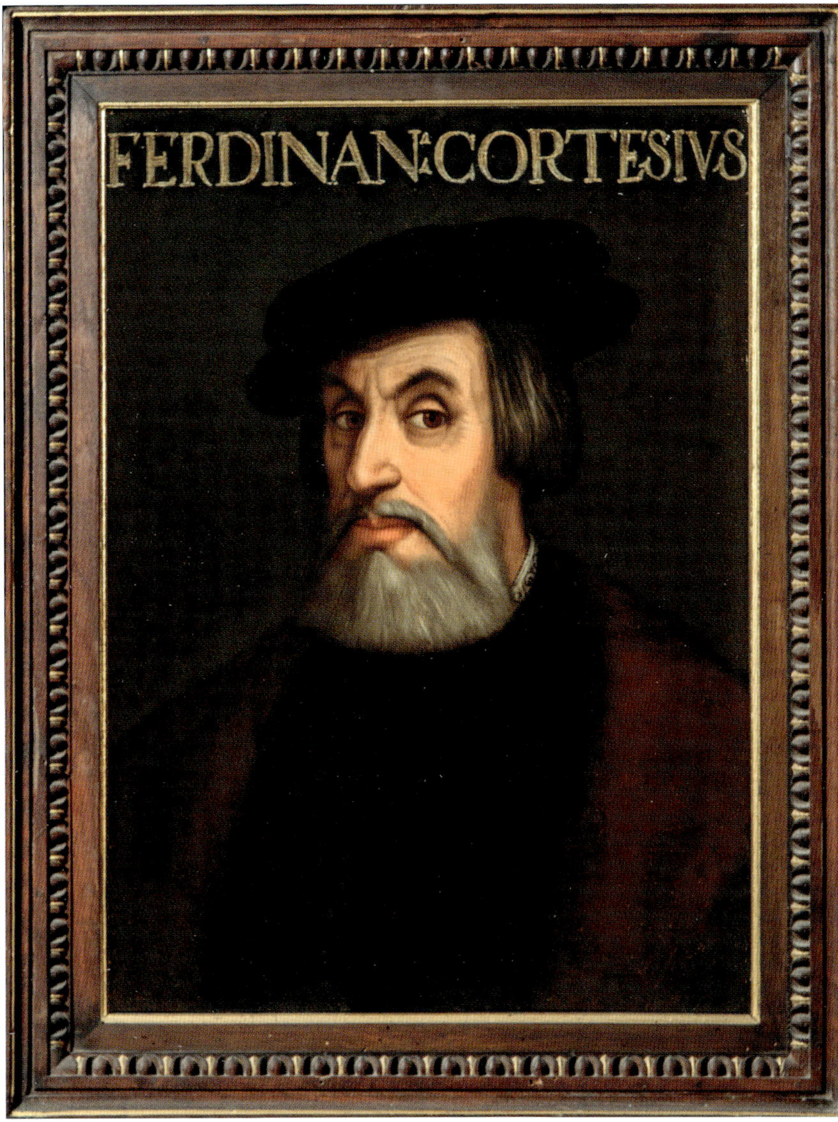
FERDINAN.^{DUS} CORTESIUS

Attributed to Johannes Vingboons and workshop
Map of Santiago Bay in Cuba

Amsterdam, *c.* 1665
Line-and-wash drawing on canvas-backed paper, 63.5 × 46.6 cm
Florence, Biblioteca Medicea Laurenziana, Carte di Castello, 40

The map depicts the bay and city of Santiago de Cuba. Santiago, on the coast of Cuba opposite the island of Santo Domingo, was founded in the summer of 1515 by the Spanish *conquistador* Diego Velázquez, who arrived in America during Columbus' second voyage. Immediately, its geographical layout made it a crucial point for expeditions along the coast of the Gulf of Mexico. The fort was a key element in a complex system of military architecture, aimed at protecting the Caribbean islands from the constant assaults by pirates.

After a couple of fortunate and less fortunate arrivals, three Spanish expeditions, financed by Diego Velázquez, the first governor of Cuba, arrived in Mesoamerica. In 1517, Francisco Hernández de Córdoba commanded the first of them: he landed in Campeche, where, on an altar, the Spaniards saw a sign in the shape of a cross, which the Maya associated with a god called Kukulcán. They thus learned of the cult of Quetzalcóatl and of the great city where, surrounded by treasures, Moctezuma II reigned, the most powerful man of those lands.

Moctezuma II was informed of the Spaniards' arrival and from then on ordered surveillance of the eastern coast. The following year, in 1518, during Juan de Grijalva's exploration, the inhabitants of the Cotaxtla river valley warned the sovereign, who sent two of his most important generals, who met with Captain Francisco de Montejo. This event represented the first meeting between representatives of the monarchs Moctezuma II and Charles V, the most powerful kingdoms of the Old and New Worlds.

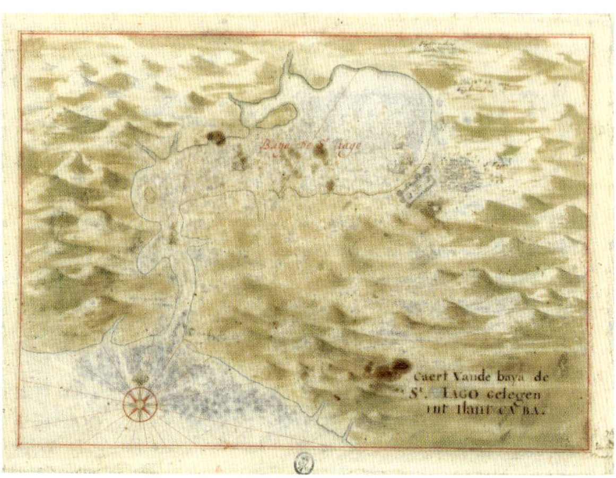

The third Spanish incursion was that of Hernán Cortés. [R.V.]

Florence, Biblioteca Medicea Laurenziana, Mediceo Palatino 220

Mexico, 16th century (1577)
fol. 406r, detail

On 18 February in the year 1519, a group of adventurers, daring to the point of risking their lives and their small fortunes, consisting of a fleet of 11 ships with 508 soldiers and 100 sailors, captained by the reckless Hernán Cortés, set sail from Cuba on an exploration mission. They landed on the island of Cozumel in February 1519. En route along the coasts of the Yucatán peninsula and the Gulf of Mexico, Cortés sporadically came across the original Maya and Chontales populations. Two encounters would prove to be of great importance: Gerónimo de Aguilar, a Spanish castaway who, after living for about eight years among the natives, had learned the Mayan language, and the gift Cortés received in Tabasco, namely an Indian woman, Malinche, who, in addition to the Mayan language, also spoke Nahuatl. Thanks to her, the conqueror was able to communicate with the inhabitants of those mysterious lands. Cortés learned of the existence of the Méxica people and their powerful ruler, Moctezuma II. On the coast of Veracruz, the Spaniards met the local Totonachi people, who informed them that there was a state that forced them to pay harsh taxes. Cortés immediately perceived the vastness and eminence of that authority and reacted by offering military protection to the Totonachi in exchange for an alliance against that power that oppressed them. [R.V.]

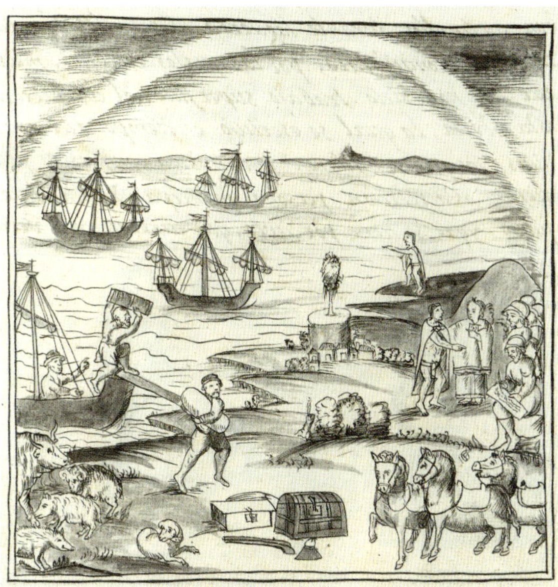

PROFECÍA

Florence, Biblioteca Medicea Laurenziana, Mediceo Palatino 220

Mexico, 16th century (1577)
fol. 425r, detail

> … como oyó la nueva Motecuzoma despachó luego gente para el recibimiento de Quetzalcóatl porque pensó que era él el que venía, porque cada día le estaban esperando y como tenia relación que Quetzalcóatl había ido por la mar hacia el oriente y los navíos venían de hacia el oriente, por esto pensaron que era él.
> ('When Moctezuma heard the news, he immediately sent people to welcome Quetzalcóatl, since he believed that it was he who was arriving, inasmuch as he was expected daily. And since they knew that Quetzalcóatl had left by sea for the East, and the ships arriving were from the East, they believed it was him').
> Bernardino de Sahagún, Mediceo Palatino 220, fol. 412v.

Through the reconstruction of the celestial cycles observed by the emperor and his wise councillors, the sovereign associated astronomical events related to Venus to the landings of the Spaniards.

Moctezuma II identified the prophecy of the return of the god Quetzalcóatl in the arrival of Hernán Cortés. [R.V.]

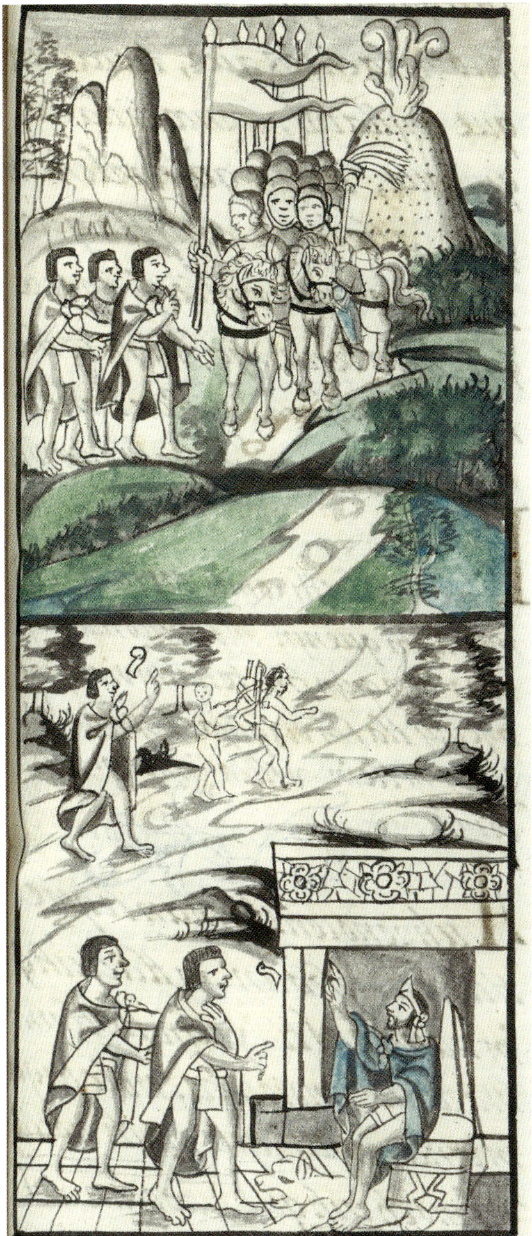

Florence, Biblioteca Medicea Laurenziana, Mediceo Palatino 220

Mexico, 16th century (1577)
fol. 415v, detail

> 'Mirad que me han dicho que ha llegado nuestro señor Quetzalcóatl Id y recebilde … Veis aquí estas joyas que le presentéis de mi parte' … Comenzaron a subir al navío por la escalera, y llevaban el presente que Motecuzoma los mandó llevar.
> ('"I was told, our lord Quetzalcóatl has arrived, go and welcome him … Here are these jewels that you will present to him from me" … They started to climb up onto the ship by the ladder, carrying the gifts that Moctezuma had ordered them to bring').
> Bernardino de Sahagún, Mediceo Palatino 220, fol. 413r.

These were Spanish ships that arrived in San Juan de Ulúa on Good Friday 1519. Moctezuma II initially attempted a peaceful approach, offering the guests lavish gifts. Being close to the coast, the Mexican emissaries arrived at Cortes' ship by canoe and offered, among other things, a snake mask decorated with turquoise mosaics. In return, Cortés ordered the feet and necks of the envoys to be clapped in irons and then had a cannon fired, causing them to faint from fright. Once they had recovered, Cortés handed them some objects, including glass beads, to take to Moctezuma II.

Cortés was drawn to all the riches that Moctezuma II had sent and, instead of choosing the option of retreating, as the Aztec emperor had expected, he ambitiously set out on the road to marvellous Tenochtitlán. On Easter Sunday, the Spaniards had already settled on some sand dunes. It is possible that they had planned their arrival in the new land between 1518 and 1520 by scheduling it on days marked by significant events that would boost the troops' morale, as they coincided with important Catholic holidays: St John's Day and Holy Cross Day. The fact that they occurred during stellar events that the Mexicans considered ominous could have been a fortuitous and significant coincidence or, more likely, brilliantly strategic planning. [R.V.]

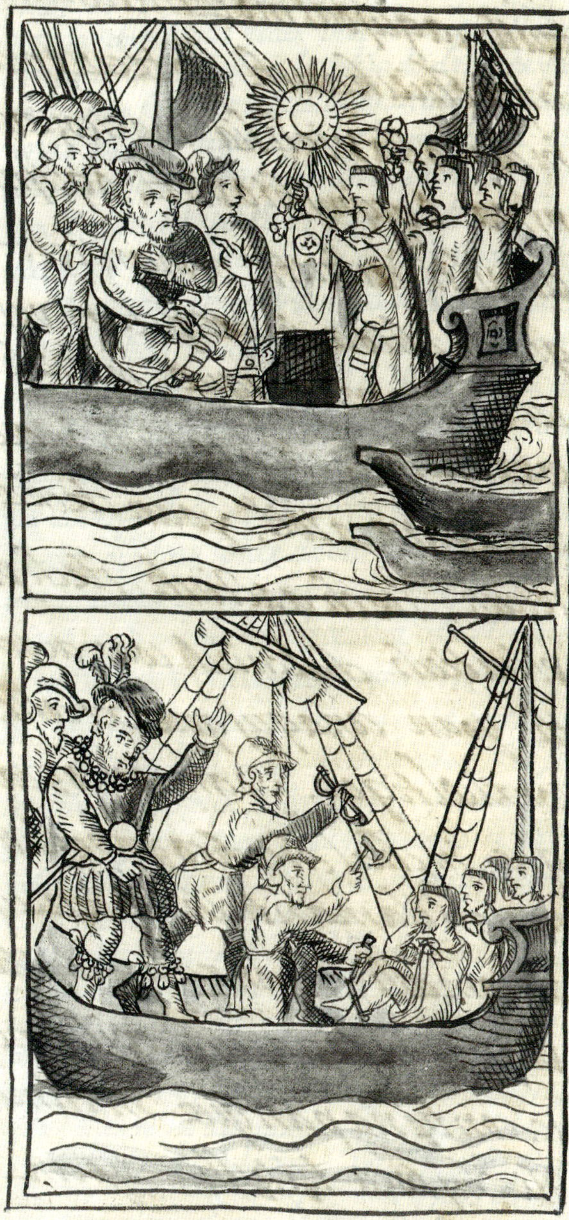

Glass beads

15th century
London, Museum of London

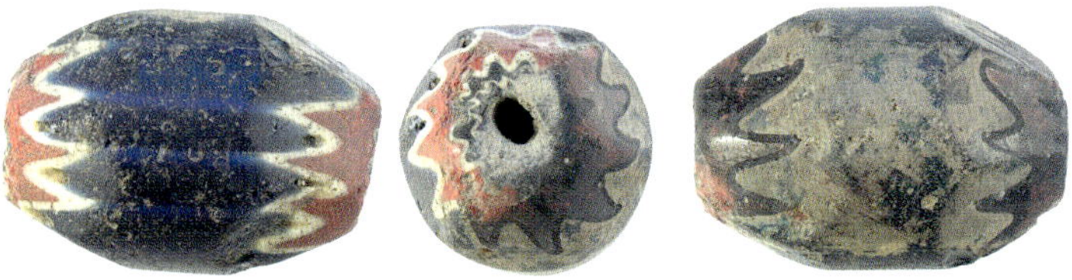

> A todos estos comunicó las nuevas que habían llegado, y mostrólos las cuentas de vidrio que habían traído los mensajeros, y díxolos: 'Paréceme que son piedras preciosas. Guárdense mucho en la recámara; no se pierda ninguna'.
> ('To everyone present, he showed the glass beads that the messengers had brought and said, "They seem to me to be precious gems; keep them diligently in the chamber, that none may be lost"').
> Bernardino de Sahagún, Mediceo Palatino 220, fol. 412v.

For a long time, Venetian glass beads had been a commodity for explorers to trade. Several Spanish chroniclers from the 16th century recount that Spaniards who arrived in the Yucatán peninsula from Cuba between 1517 and 1518 traded glass beads with Mesoamericans in exchange for gold jewellery, blankets, clothes and food. Many of them were transparent and had various colours inside, but the ones the natives liked best were the green or blue ones.

In the 17th century, among some populations of Africa, such as the Ashanti of Ghana, glass beads were considered rarer than semi-precious stones and precious gems, and were therefore highly coveted. It became customary to barter them for slaves to be shipped to the New World or for gold and ivory, which reached the European courts. The beads used to buy slaves were biconical or cylindrical, faceted, and almost all blue. They are still called 'Slave Beads'. [R.V.]

OBSIDIAN

Indigenous art,
Snake-shaped figure
Chichenitzá, 2013
Obsidian, 7 × 2.5 cm
Florence, private collection

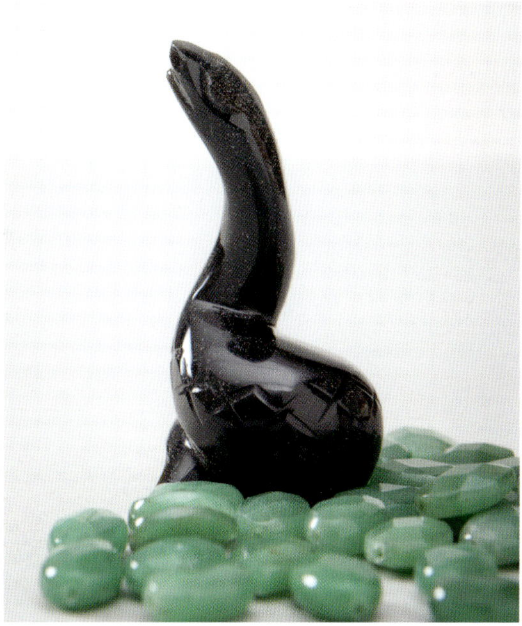

Upon their arrival, in the territory where present-day Mexico is located, the Spanish found a highly militarised and hierarchised society, maintained by the subjugated peoples.

The Mexica culture excelled in military aspects thanks to the alliance between the dominant rulers of Tenochtitlán, Texcoco (Tzcóatl) and Tacuba (Cuauhtlatoa), united in the 'Triple Alianza' or the so-called Confederation of the Anáhuac, the origin of the Mexica army.

Surprisingly, this empire subjugated Mesoamerica using weapons typical of the Neolithic period: spears, arrows and knives made of a volcanic rock called obsidian, which was found in abundance in the region. Obsidian is an igneous rock, in other words, formed when volcanic magma cools and solidifies on the surface. It is hard and brittle at the same time, and obsidian blades are extremely sharp. The obsidian arrowheads of the Mexica were feared by the *conquistadores* because they caused serious damage. Once they penetrated the body, it was almost impossible to extract them. Bernal Díaz del Castillo (1492–1584) called the Mexican warriors 'indios flecheros'. [R.V.]

Florence, Biblioteca Medicea Laurenziana, Mediceo Palatino 220

Mexico, 16th century (1577)
fol. 433r, detail

> '¡Oh, señor nuestro! Seáis muy bien venido. Habéis llegado a vuestra tierra, a vuestro pueblo y a vuestra casa, México' … Como don Hernando Cortés hubo entendido lo que había dicho Motecuzoma, dixo a Marina: 'Decilde a Motecuzoma que se consuele y huelque, y no haya temor, que yo le quiero mucho y todos los que conmigo vienen. De nadie recibirá daño'. ('"Oh, dear Lord! You are most welcome. You have arrived in your land, amongst your people and in your home, in Mexico" … When Don Hernando Cortés understood what Moctezuma had said, he said to Marina: "Tell him to take comfort and rest assured, and not to be afraid, that I and everyone with me love him very much. No one will harm him"').
> Bernardino de Sahagún, Mediceo Palatino 220, fols. 431v, 432v.

The meeting between Moctezuma II and Cortés finally took place. On 8 November 1519, eight months after landing in Cozumel, Moctezuma II arrived at the rendezvous accompanied by a number of councillors, clerics and aristocrats. He was adorned with quetzal feathers. His robes were embroidered with precious stones, pearls and exotic plumage. His subjects rolled out fine carpets so that the emperor would not touch the ground, as chronicler Díaz del Castillo recounts, astonished by the extraordinary sumptuousness of it all:

> The great Montezuma was about forty years old, he was of good stature and well-proportioned, with a supple and fairly lean build, and his colour was not very dark, but was that of an Indian, and his hair was not very long, but covered his ears… and his face was somewhat elongated and cheerful, and his eyes were good, and he showed a great deal of love in his gaze and, when necessary, a great deal of seriousness. He was very neat and clean, he bathed once a day every afternoon.
> Díaz del Castillo, *Historia verdadera de la conquista de la Nueva España* [1632], ed. Mexico, 1993, pp. 248–9.

Hernán Cortés was aware that conquest was imminent and inevitable and wished to be a protagonist in this process. However, the success of this enterprise would never have been possible without the alliances forged with the other peoples subjugated by the Mexican empire, as well as the crucial presence of Malinche. Not only did she speak the Nahuatl and Mayan languages, but she also perfectly understood the mentality of the Nahua peoples and the influence of México-Tenochtitlán. For this reason, in the depiction of the meeting between Moctezuma II and Cortés that took place at the site of Huitzilan ('place of the hummingbirds'), the central figure in the Florentine Codex is the Nahua woman,

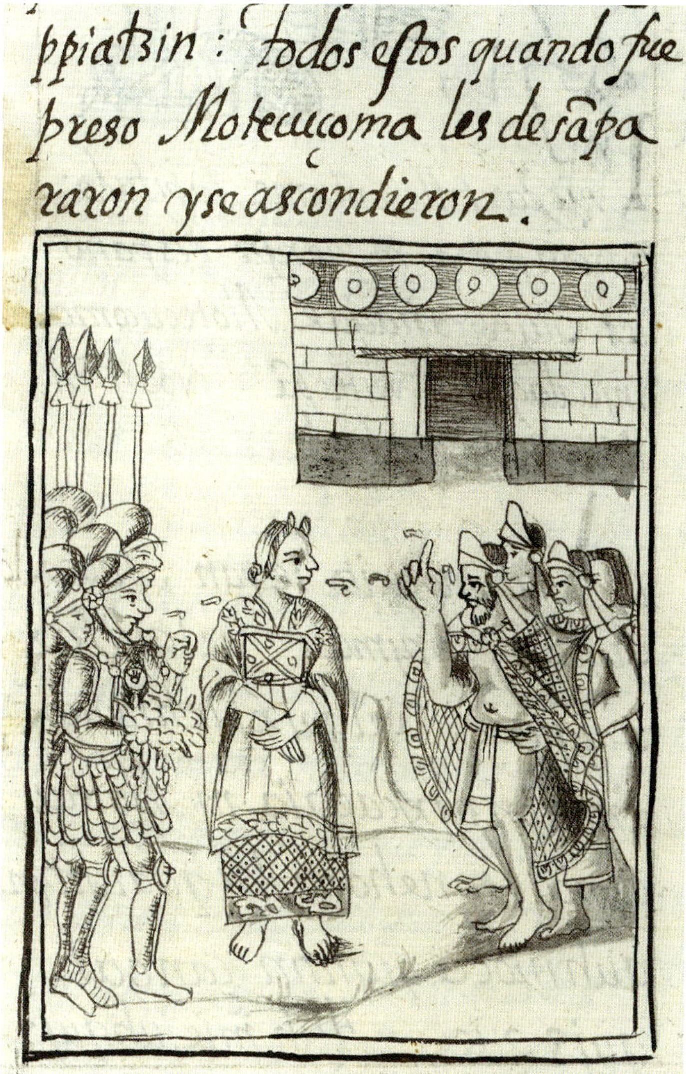

Malinche (also known as Malintzin). It is she who represents the meeting point between the two conflicting powers and the fulcrum of the alliance in favour of the newcomers from overseas. [R.V.]

Florence, Biblioteca Medicea Laurenziana, Mediceo Palatino 220

Mexico, 16th century (1577)
fol. 419v, detail

Moctezuma II invited Cortés to the heart of the city and hosted him in the palace of Axayácatl. The emperor, who loved of luxury and banquets, treated his guests to a sumptuous meal that neither Cortés nor the soldiers in his retinue could ever have imagined. At the end of the meal, according to Bernal Díaz del Castillo, a dense, flavourful beverage was served in golden cups, enjoyed by all the diners, and particularly the emperor himself. The *conquistadores* were the first westerners to taste chocolate, *cacaoatl.* Treated as distinguished guests, and surrounded by such splendour, they felt that the fortune that they had hoped for was finally at hand. The Mexica people gave Cortés the finest welcome, but with fatal consequences: without knowing it, they offered their most prized drink to the man who would be mainly responsible for their demise. [R.V.]

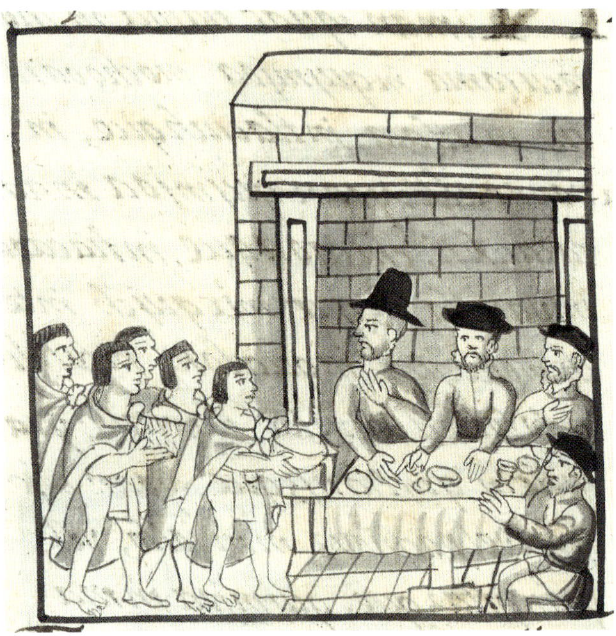

Florence, Biblioteca Medicea Laurenziana, Mediceo Palatino 220

Mexico, 16th century (1577)
fol. 408r, details

The encounter turned into a brutal clash of civilisations. Hernán Cortés, aged just 34, turned a simple exploration mission into a war of conquest.

Once he reached the lodgings prepared for him and his men, he seized the treasure stored in the palace and, even worse, imprisoned Moctezuma II and the ruler of Tlatelolco, Iztcuauhtzin.

The emperor then saw his power and political relevance decline. His actions were limited to demanding food supplies for the Spaniards. He was unable to prevent them from grabbing the treasures from palaces and temples. According to the chronicles of the time, Moctezuma II tried to calm the Mexicans after the massacre in the Templo Mayor by Pedro de Alvarado, who had unleashed a general insurrection. It was this act that led to the death of the Mexica emperor. There are two versions of the magnicide: the Spanish version claims that Moctezuma II died when a stone was hurled at his temple by his own people, while the indigenous version states that he was killed by the Spaniards. Regardless of how he died, the insurrection was only quelled by something new and unexpected: smallpox, a disease unknown to the natives. The population was decimated. Finally, on 15 August 1521, Cuauhtémoc, the last *huey tlatoani* (emperor) of Mexico, was captured, and with him the great Tenochtitlán fell.

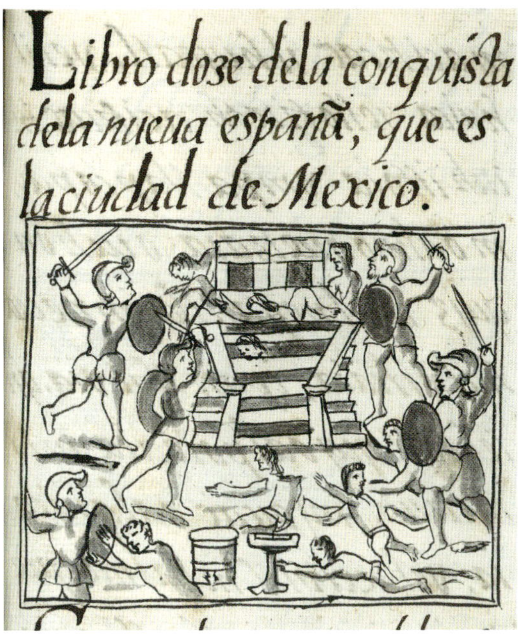

However, the razed city did not disappear completely. In the third *Cartas de Relación a Carlos V*, dated 15 May 1522, Cortés informed the king that he had taken steps to rebuild it:

> Considering, moreover, that the city of Temixtán [Teno-

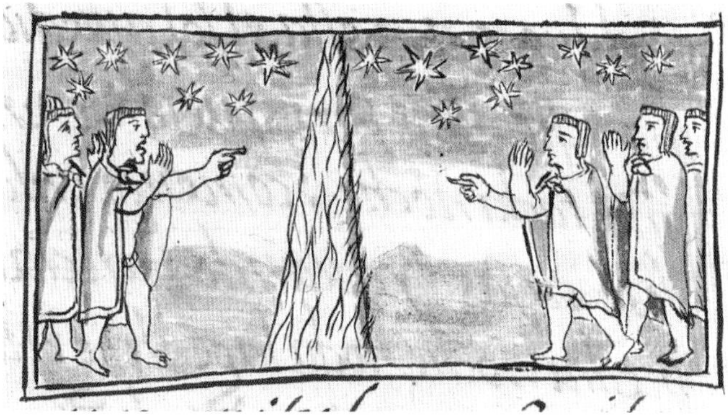

> chtitlán], which had been completely destroyed, had always been so famous and prestigious, it seemed fitting to rebuild it as our city … the reconstruction of Temixtán has been very successful, and believe it, Your Majesty, that every day the city will become more beautiful, so that, just as before it was the queen and capital of all these provinces, so it will be from now on.

According to a Mayan legend, from then on the quetzal stopped singing its usual melodies, and would only begin again on the day its lands regained their lost freedom. [R.V.]

HUICHOL

Huichol art, *Curandera e hijo*

Wirikuta, San Luis Potosí, 2021
Multi-coloured cotton yarn on wooden board (wewiya), 35 × 35 cm
Florence, private collection

During the Spanish conquest, some peoples managed to escape to the highlands of the western Sierra Madres, a mountain range in present-day western Mexico. Since the Sierra was virtually inaccessible, it could not be conquered. The indigenous Coras, Huicholes and Tepehuanes of the Sierra del Nayar resisted for at least two centuries (1521–1721) without submitting to the Spanish. Among the survivors of the conquest and genocide, the members of the Wixárika culture, commonly known as the Huichol, are today custodians of traditions that they have managed to preserve over the centuries. They continue to worship their ancient gods and their sacred and powerful peyote, an entheogenic substance used in their religious and spiritual practices. They have been great embroiderers, great shamans and myth-tellers for generations. [R.V.]

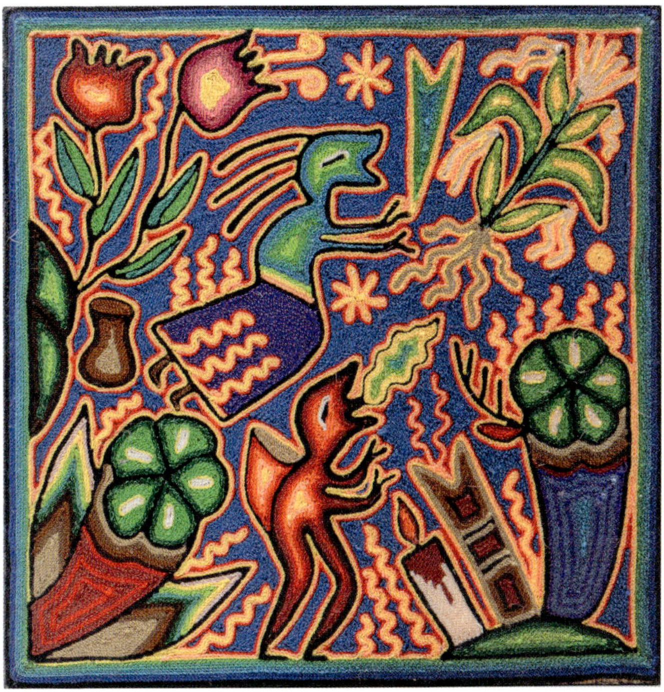

Map of Mexico City portrayed immediately after the Spanish conquest

in Hernán Cortés, *Segunda y tercera cartas de relacion*, Nuremberg, F. Peypus, 1524

Tenochtitlán refounded under the name 'México, Regia et celebris hispaniae novae' became capital of the viceroyalty or kingdom of New Spain.

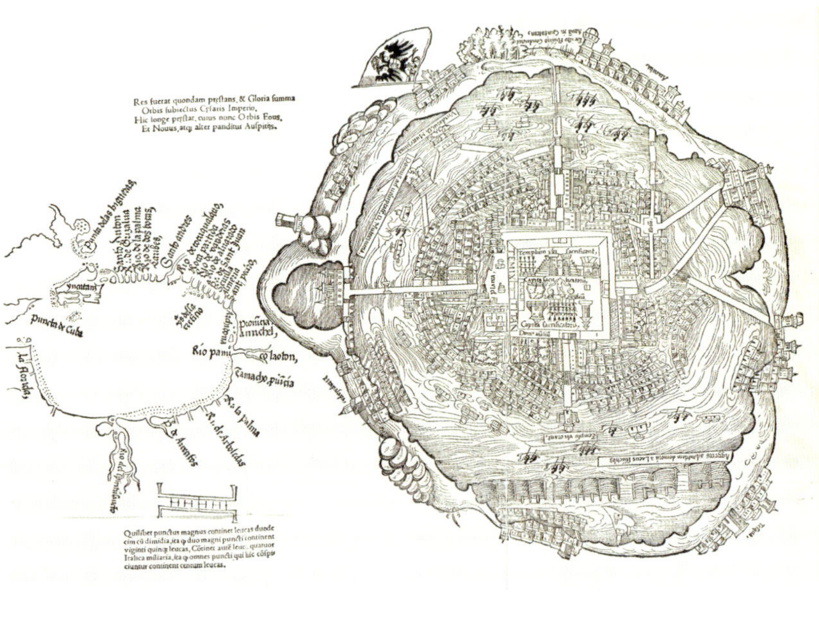

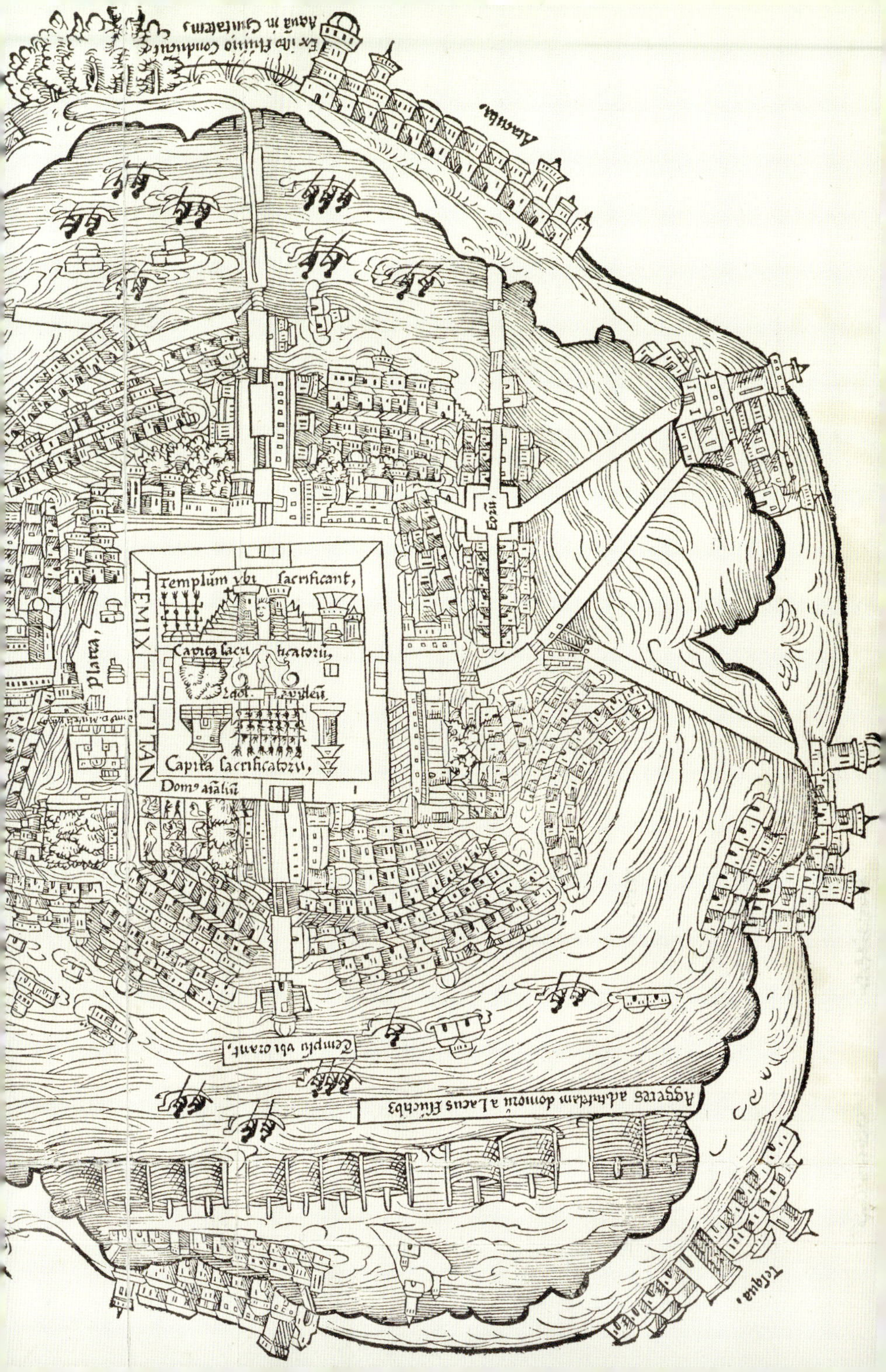

Templum ubi sacrificant,
TEMIX
TITAN
Platea,
Capita sacrificatorii,
Capita sacrificatorii,
Dom° aīaliū
Templu ubi orant.

3. PECUNIA

Book VIII, Noblemen and rulers
'de los reyes y señores y de la manera que tenían en sus electiones y en el gobierno de sus reinos'

Book IX, Trade, merchants and artisans
'de los mercaderes, oficiales de oro y piedras preciosas y pluma rica'

MÉXICO

Attributed to Johannes Vingboons and workshop, after Juan Gómez de Trasmonte, 1628

Plan and localisation of Mexico City

Amsterdam, *c.* 1640–68
Line-and-wash drawing on canvas-backed paper, 60.2 × 47.5 cm
Florence, Biblioteca Medicea Laurenziana, Carte di Castello, 43

The conquest of the 'wealthiest and most fascinating part of the world', according to Montaigne, was made possible not only by the alliances of the *conquistadores* with the peoples subjugated by the Aztecs, but also by the use of firearms, steel, means of transport (the horse and wheeled vehicles), as well as the diseases that the Spaniards brought with them from Europe. After the defeat of the Mexican Empire, the conquered lands became the domain of the King of Spain and Emperor of the Holy Roman Empire, Charles V of Habsburg.

The capital was transformed into a large colonial city. The urban structure of the new city, however, followed that of the Aztec city, based on a grid plan. Maps of Mexico City prior to the 18th century are very rare. Juan Gómez de Trasmonte (1595–1647) is the author of a map made in 1628, practically a detailed perspective of the city. It is significant because it was made shortly before the catastrophic flood of 1629, which caused a considerable rise in Lake Texcoco on which the city was built, and because it is the only one that gives a complete view, with streets, gardens, lakes, houses and the most important buildings, including convents, monasteries, hospitals, schools, churches and monumental buildings in the main square. [R.V.]

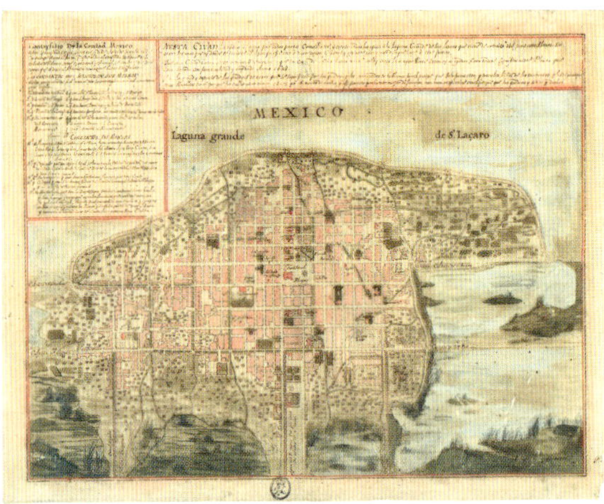

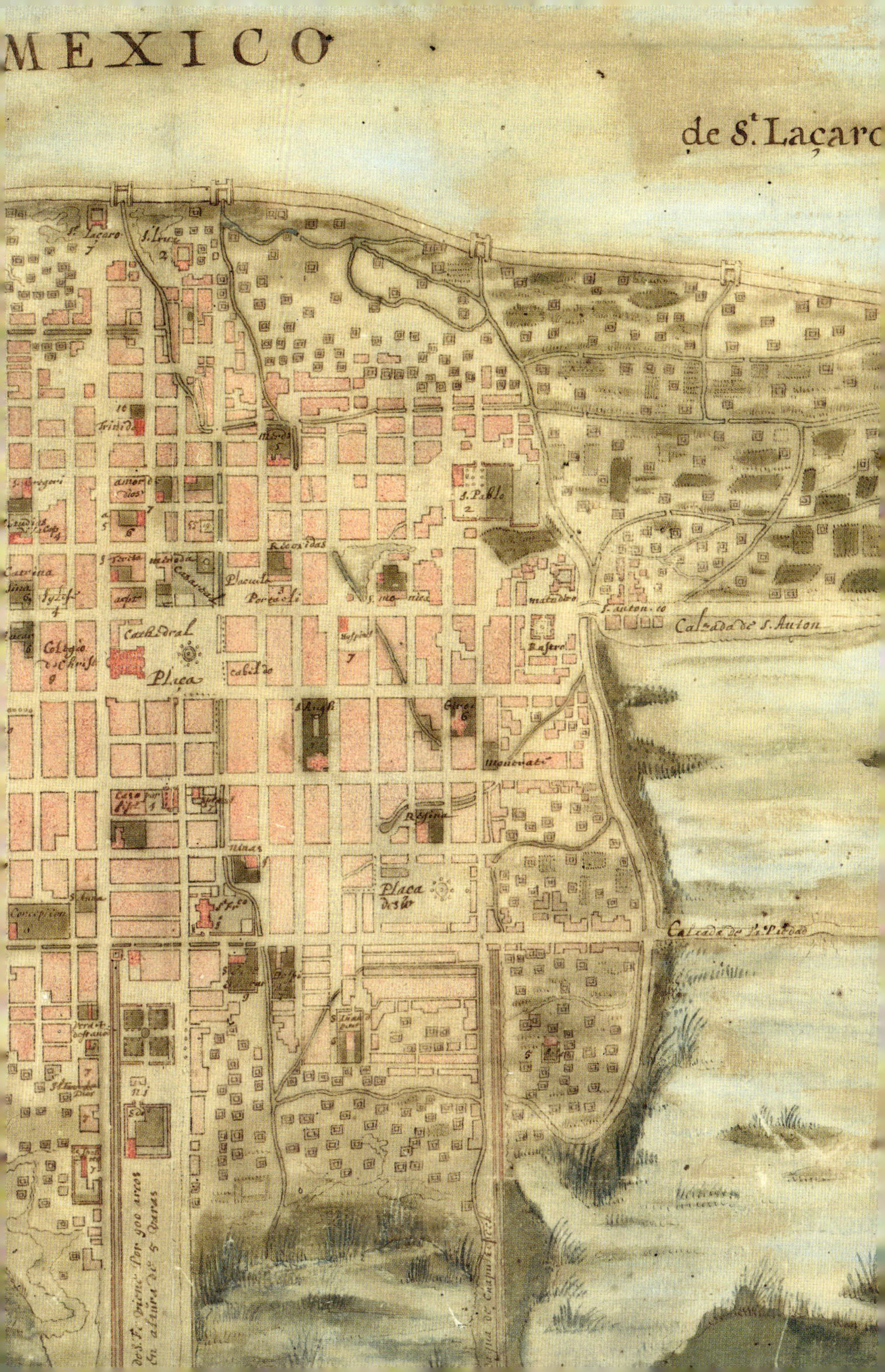
MEXICO
de S.t Laçaro
S. Pablo
Recogidas
Cathedral
Plaça
Calçada de S. Anton
Calçada de la Piedad
Placa

MOCTEZUMA'S TREASURE

Florence, Biblioteca Medicea Laurenziana, Mediceo Palatino 220

Mexico, 16th century (1577)
fol. 434v, detail

> … comenzaron a preguntar a Motecuzoma por el tesoro real, para que dixese dónde estaba. y él los llevó a una sala que se llamaba teucalco, donde tenían todos los plumajes ricos y otras joyas … [y por] quitar el oro destruyeron todos los plumajes y joyas ricas. Y el oro fundiéronlo y hicieron barretas.
> ('They began to interrogate Moctezuma about the royal treasure, asking him to reveal its whereabouts. He led them to a room called a *teucalco* … then they began to remove the gold from the feathers, shields and ceremonial decorations and, in the process, destroyed all the feathers and melted the gold down into ingots').
> Bernardino de Sahagún, Mediceo Palatino 220, fol. 434v.

In order to maximise the cargo of treasures on board Spanish ships, the precious pieces of finely worked gold were melted down and turned into ingots to take up less space. From the newly conquered and colonised colonies, ships laden with riches arrived in Spain.

Charles V was entitled to a tribute from the spoils of war, as well as products and goods from the new lands. [R.V.]

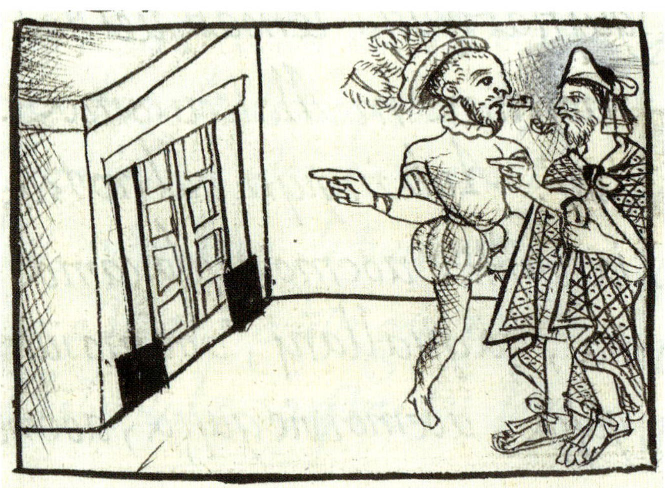

QUINTO REAL

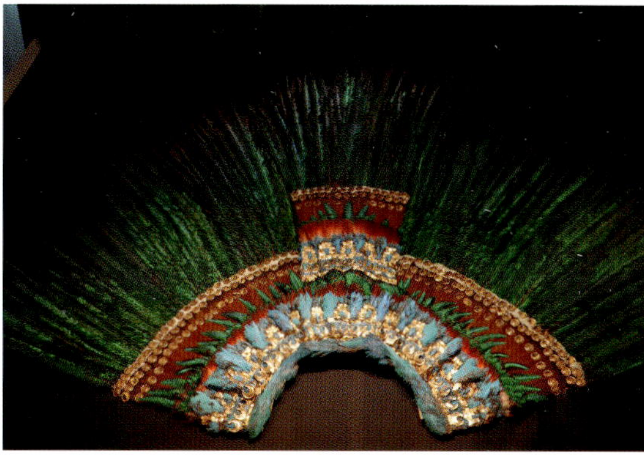

Mexica craftsmanship, ***Quetzal feather headdress***
16th century
Vienna, Weltmuseum

The Quinto Real was the main tax in the New World. It was a tribute owed to the Real Hacienda equal to one fifth of the value of goods, including the mining and production of precious metals. The tax was also levied on spoils of war and even on temple and cemetery treasures unearthed by the conquerors. The system also provided for the taxation of barter-based trade through another tax called the Quinto de Rescates.

Along with the first shipment of the Quinto Real, the young Charles V was gratified by the huge gifts sent by Hernán Cortés. Cortés' trusted emissaries, who arrived in Spain in October 1519, reached the emperor in Tordesillas and delivered what was entrusted to them, arousing great interest at court. Pedro Mártir de Anglería (1457–1526) himself, who was present at the event, expressed these words: 'It seems to me that never have I seen anything like this, which, because of its beauty, is able to draw the gaze of men so intensely'. Some of the relics continued their journey to Flanders, the young ruler's homeland, and their magnificence was such that even Albrecht Dürer (1471–1528) was impressed: 'they were capable of melting and modelling precious metals, engraving the hardest stones, and meticulously composing mosaics of multicoloured feathers, which decorated the shields, the leaders' capes and the images of the gods'.

The second cargo of Quinto Real met with an unfortunate fate. The ocean crossing necessary to travel from the Old World to the New World and back was full of perils, mainly storms and raids by privateers. Numerous cargoes were lost and among them the second cargo, plundered in favour of the King of France. The various hypotheses on the identity of the corsair commander involved in this theft include one naming Giovanni da Verrazzano, a Florentine explorer and navigator, as the person responsible. Charles V's protests to Francis I were vociferous, claiming that there was no formal state of war between the two countries. However, the French sovereign withheld the spoils, mockingly replying to Charles V, 'Show the testament of our father Adam in which all those lands were bequeathed to Your Majesty'.

Centuries later, the expression 'me quedé sin un quinto' still exists in Mexico, meaning 'I've run out of money'. [R.V.]

Colonial art, *Stoup*

Mexico, 18th century
Silver, 35.5 × 25.5 cm
Florence, Lourdes Echavarría Collection

Annually, America sent approximately 67 tonnes of gold and 1200 tonnes of silver to Spain, equivalent to more than 250,000 ducats. From the port of Veracruz, the gold and silver travelled to the port of Seville and from there to Toledo, where it arrived in the hands of Emperor Charles V, who distributed it between the expenses of war and those of the imperial court, throughout his vast domains. [R.V.]

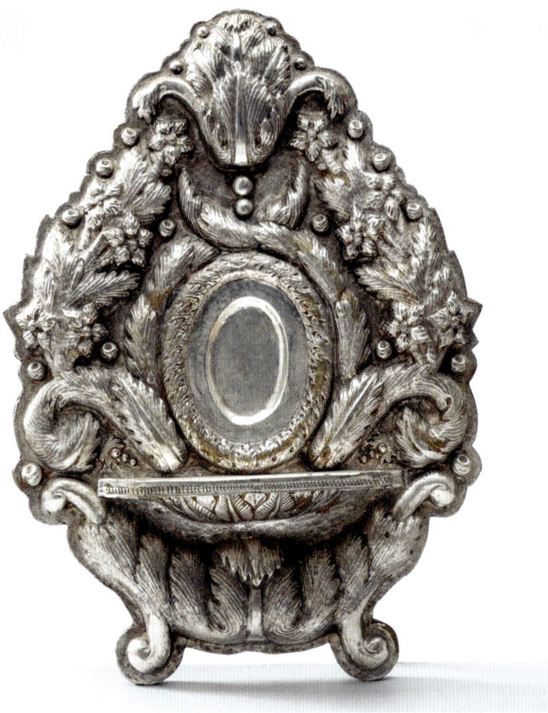

GOLD

Not everyone knows that part of the first shipment of gold stolen from Mexico is today in Italy, in the papal basilica of Santa Maria Maggiore in Rome. Founded around the year 360, the basilica was rebuilt by Pope Sixtus III (432–40) and decorated for the first time with wonderful gold mosaics. This is the first figurative cycle documented in a Roman church. Enriched over the centuries with new and ever more majestic decorations, thanks to continuous interventions such as those by Pope Eugenius III (1145–53) and Pope Nicholas IV (1288–92), the basilica survived into the 15th century as a legacy of venerable antiquity, rendered sacred and hieratic precisely by the tiles of its impressive mosaics in the apse and on the façade. When the future pope Rodrigo Borgia (1431–1503) – at the time still a cardinal – commissioned a new coffered ceiling from Lorenzo the Magnificent's favourite architect, the extraordinary Giuliano da Sangallo, he expressly requested that the wooden elements, carved around the highly recognisable heraldic coat of arms of his family, be covered with a precious layer of gold leaf. Cardinal Borgia, who finally ascended to the papal throne in 1492 with the name of Alexander VI, obtained a gift from the Spanish royal family that allowed him to complete the prestigious work: a load of pure gold from the New World. [L.H.]

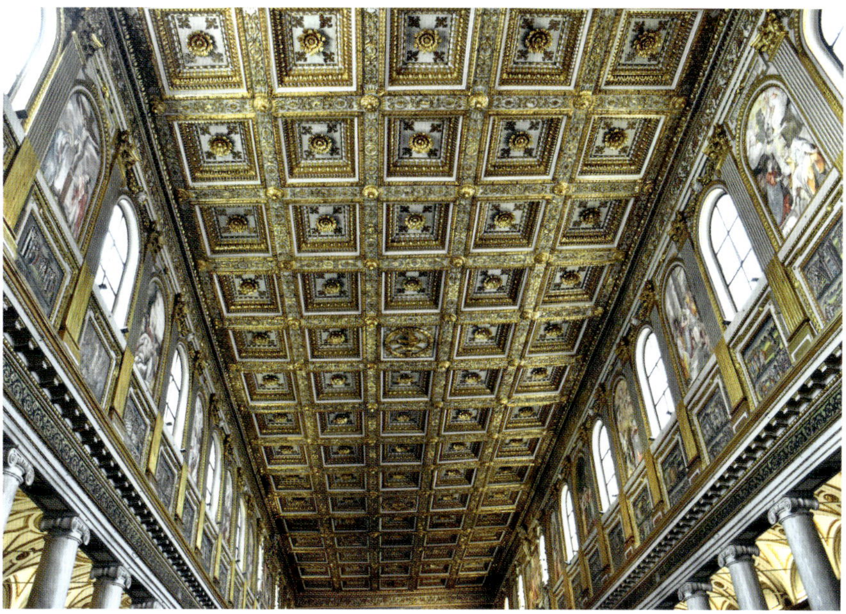

Rome, Basilica of Santa Maria Maggiore, detail of the ceiling.

Commemorative coin of the 500th anniversary of historic memory of México-Tenochtitlán

Casa de Moneda de México, 2021
Nickel silver, bronze, nickel, zinc, aluminium, ø 30 mm
Florence, private collection

At the dawn of the turbulent years of the conquest, the Cédula Real of 14 September 1519 authorised Hernán Cortés to coin, melt and brand the ornate gold accumulated in large quantities in Mexico's wealthy Tenochtitlán, using the palace of Axayácatl for this purpose. In this way, what can be considered as the first mint or 'Casa de Moneda' in the regions of the so-called New Spain was established. With the help of the skilled and experienced silversmiths of Azcapotzalco, it provided the Castilians with the necessary practice to later carry out more elaborate goldsmithing. In 1521, when Cortés settled in Coyoacán, in the southern part of present-day Mexico City, prompted by both royal and popular pressure, he quickly arranged for a new 'Casa de Moneda' to be established there, for which he adapted one of his rooms. In the presence of three crown officials, he proceeded to melt down what, according to the documentation collected by Francisco Antonio de Lorenzana, amounted to 'more than one hundred and thirty thousand castellans' – the main unit at the time – a fifth of which was sent to the King of Spain.

Pictured here is the commemorative coin issued by the Casa de Moneda de México to mark 500 years of México-Tenochtitlán's historical memory. On the front, in the centre, is the national coat of arms in sculptural relief and on the reverse, creating a reflection, the Metropolitan Cathedral, the Templo Mayor as a latent image and the text 'Fusión Cultural'. The years '2021' and '1521' are to the right of the face value, which corresponds to '$20'. The 20 pesos coins are those produced in the smallest quantities and are often chosen to commemorate important events in the history of Mexico, which is why these specimens on display are extremely rare and highly coveted by numismatists, ready to pay extravagant sums for them. [R.V.]

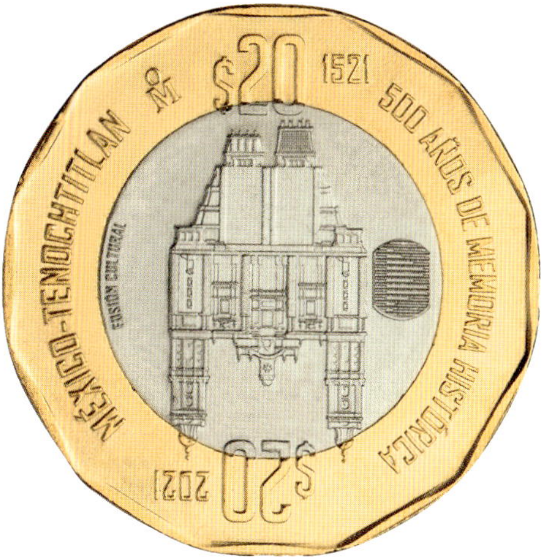
$20
1521
500 AÑOS DE MEMORIA HISTÓRICA
MÉXICO-TENOCHTITLAN
FUSIÓN CULTURAL
2021

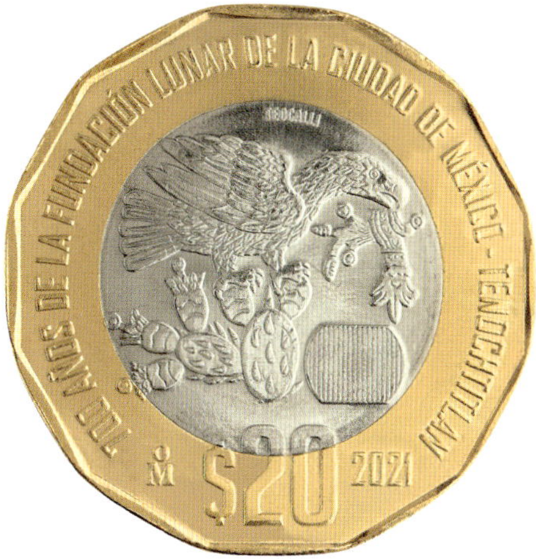
700 AÑOS DE LA FUNDACIÓN LUNAR DE LA CIUDAD DE MÉXICO - TENOCHTITLAN
$20
2021

OAXACA

Florence, Biblioteca Medicea Laurenziana, Mediceo Palatino 220

Mexico, 16th century (1577)
fol. 275r

In Mesoamerica, the conception of currency had until then been very different. Cortés stood out among the first Europeans for his interest in the indigenous custom of adopting cocoa as a form of currency. In the *Segunda Carta de Relación a Carlos V*, dated 1520, he describes cocoa as 'a fruit like almonds that they sell in pulverized form and hold in such esteem that they exchange it as currency and with it buy everything they need in markets and other places'. Cortés himself sometimes paid his soldiers with cocoa. In another letter to the emperor, dated 1524, he notes that the value of a copper axe – equivalent to an ingot of metal – roughly corresponds to 8000 cacao beans.

Cortés received an area with thousands of cacao plantations in payment for his services to the Crown in 1529. The young ruler Charles V granted Cortés a noble title and domain, perpetual and hereditary, the origin of the Marquisate of the Valley of Oaxaca. Rather than being a contiguous unified territory, it was made up of twenty-two cities, which formed seven territorial portions corresponding to the current territories of Oaxaca, Estado de México, Morelos, Veracruz and Michoacán. In the beautiful city of Cuernavaca he built his castle-palace and established the seat of his marquisate, exploiting the natural environment to generate wealth. In place of the king, the marquis administered justice, so that he became the immediate authority. The Marquisate endured for three centuries, from its creation until Mexico gained independence from Spain in 1821. With the exception of brief periods during which it was confiscated by the Crown, it remained in the hands of the descendants of Cortés.

Oaxaca today is traditionally the region of Mexico with the highest consumption of cocoa, and chocolate, both solid and liquid. [L.H.–R.V.]

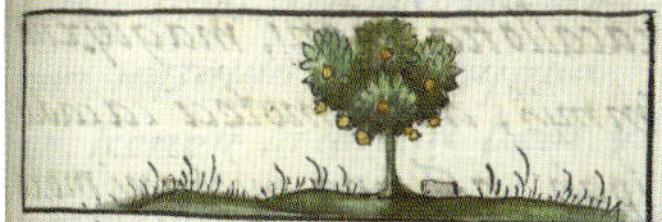

¶ Al arbol donde se hase el cacao
llaman, cacaoaquavitl: tiene
las hoias anchas, y es acopado,
es mediano el fruto que hase,
es como maçorcas de mayz, o po
co mayores, y tienen de dentro
los granos de cacao, de fuera es
morado, y de dentro encarnado
o bermejo. quando es nuevo si se
bebe mucho enborracha, y si
se bebe templadamente refrige
ra, y refresca.

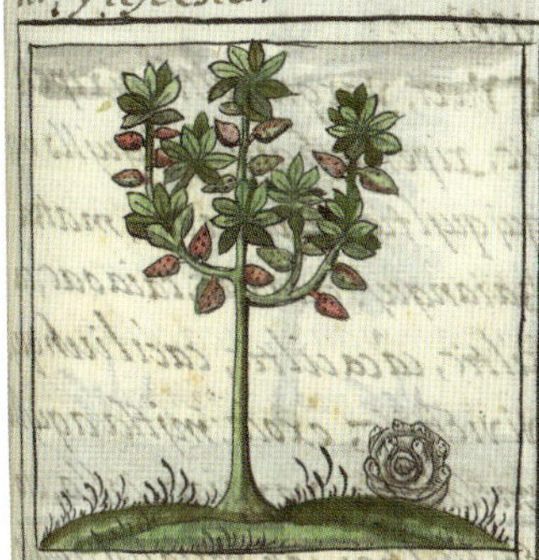

¶ Ay vnos arboles que se lla
man, teunacaztli, o teonacaz
quavitl, o vey nacaztli: las flo
res deste arbol, son muy aro
maticas y preciosas, y tienen
fuerte olor, y son muy ama

viti, teiztlacmeiaлti. njxococivi,
noztlacmeia. noxocolia, njxoco
civi.

¶ Cacaoaquavitl, mapapatla,
oac, çanquauhiaoalli: injtla
aquijllo iuhquijncintli, iuhqn
elotl: motocaiotia cacaoacin
tli, cequj tlapalcamjltic, cequj
azcamjltic, cequj tetexocamjl
tic: injiollo injiticca, injitic
tenticac, iuhquijn tlaolli: ieieh
in nenj, injtvca cacaoatl: inin
qualonj, yoanj. Inin cacaoatl
in mjec nj, in mjec ioa: occenca
ieh in xoxouhquj, in amanecoa
teivinti, tetechquj3, tequaivin
ti, teixmalacacho, teiollopolo,
teiollo tlavelilotili: in çan vel
ipan mjh tecelti, tececelti, teiol
lali, teacotlaz; iemjtva. njno
cacaoavia, njnotenciaoa, nj
nocecelia.

¶ Teunacaztli, teunacazquavitl,
veinacaztli: injtla aquijllo, velic
aviac: injiaca motetzaoaca ine
cuj, teiacaxelo, chicaoac: coz
tic, tomjiollo, tomjlollo, ihca

– Here we are, – said the Fox. – Now get down and dig a little hole with your hands, to put your gold pieces in.
Pinocchio obeyed. He dug the hole, put in the four gold pieces he had left, and covered them up.
– Now, then, – said the Fox, – go to the milldam close by, get a bucket of water, and water the ground where you have sown your seed.
Pinocchio went to the mill-dam, and, since he did not have a bucket, he took off one of his shoes, filled it with water, and watered the earth which covered his money.
– What else must I do? –
– Nothing more, – replied the Fox. – Now you can go away. Come back in about twenty minutes, and you will find a little tree already growing, with its branches covered with money.
The poor marionette was almost beside himself with joy. He thanked the Fox and the Cat a thousand times, and promised them a fine present.
– We don't want presents, – replied those two miserable wretches; – it is enough for us to have shown you how to get rich without work. That makes us perfectly happy.
With these words they wished Pinocchio a splendid harvest, bade him farewell, and went about their business.

Carlo Collodi, *The Adventures of Pinocchio*, Chapter 18, Eng. trans. London–Glasgow, n.d., pp. 102–3.

In the Mexica economy, the official medium of barter for transactions was the cacao bean, the plant of which was deemed to be sacred. It could therefore truly be said that 'money grew on trees' and that one's quality of life was directly proportional to the number of cacao beans possessed.

In 1526, the historian Peter Martyr d'Anghiera called the beans *Amygdala Pecuniaria* (monetary almonds), specifying that it was a blessed currency, since, as it could neither be buried nor kept for long, it did not encourage usury or avarice. His work *De orbe novo* contains a wonderful observation on the subject:

> But it is imperative to note here that money grows on trees. For they have money that, in order to obtain it, it is not necessary to open the viscera of the earth or turn to the avaricious men who possess it and covet it more than anything else; nor does one run the risk of losing it in war or returning to the depths of the earth and the caverns of mother earth, as one does with gold and silver. It is not risky money, because it grows on trees.

It may come as a surprise that in an advanced civilisation such as the Mexica there was no better medium of exchange than cacao, but if one compares it with other currencies,

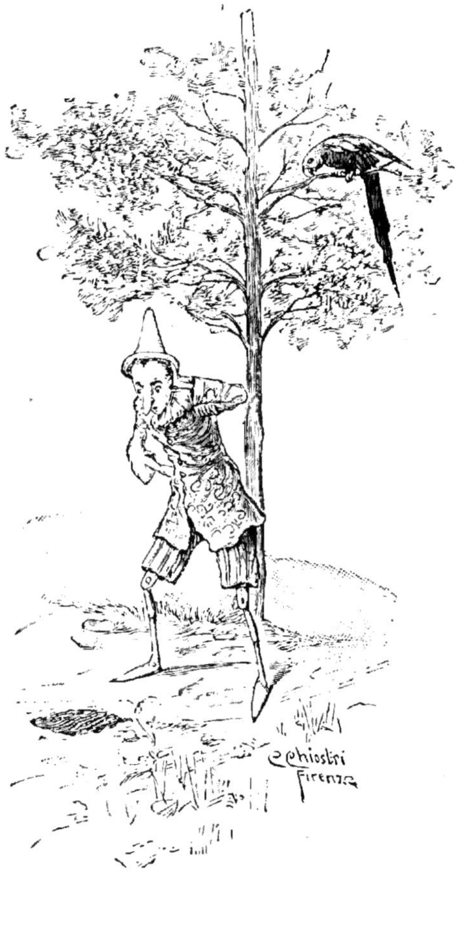

Carlo Chiostri, illustration from Carlo Collodi, *The Adventures of Pinocchio*, published in Florence in 1902.

one will see that this bean was superior to them, since in addition to being used as currency, cacao was also used to prepare drinks, with a strong symbolic significance, whose consumption was exclusive to the ruling classes:

> Yet each time an Aztec took a drink of chocolate from the brightly painted gourd cups in which it was served, he was, so to speak, drinking real money; the only equivalent to this in our own culture would be to light one's cigar with a twenty-dollar bill. Small wonder that its consumption was the prerogative of the elite.
> Sophie D. Coe, Michael D. Coe, *The True History of Chocolate*, London, 1996, p. 99. [R.V.]

FLORIN

Gold florin, Alberese hoard

15th century
Sheet of gold cut and minted, ø 2.1 cm
Florence, Museo Archeologico Nazionale, inv. 89564

From the time of the fall of the Roman empire, European states constantly suffered from a widespread shortage of gold, which culminated in Charlemagne's monetary reform which, for about four centuries, made silver the lynchpin of the entire currency system. While the Arab world, through the dinar, continued to mint gold coins, in Europe only *denarii*, pence, *grossi*, etc. were created. Then, with the reopening of important trade routes with Africa – starting in the 12th century – the first signs of a change began to appear. Amalfi coined the *tarì*, the Kingdom of Sicily the *augustale*: gold was beginning to circulate again. But, of all the new coins created at this time, one, in particular, marked an epochal breakthrough. And it was born in Florence: the florin. The florin was minted for the first time in 1252, under the *podestà* Filippo Ugoni. Its name, of course, derives from the flower represented on the obverse of the coin. It was a symbol of the growing banking power of Florence and an expression of a city that was experiencing prodigious commercial expansion. Made of pure 24-carat gold, the florin had a very high purchasing power. The high denomination and constant weight of the coin made it a much-appreciated means of payment, destined to be imitated by many European states (Britain, France, Switzerland, the Netherlands) and other powerful Italian municipalities (Venice and Genoa).

While in the Mexica empire cacao was the strongest trading currency, at the same time in the Old World the florin was the most important currency, issued until the fall of the Republic of Florence in 1533, when it was replaced by the Medici *scudo*. [L.H.]

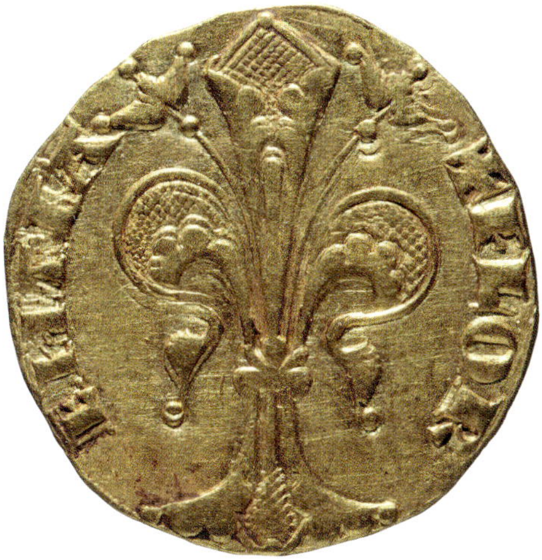

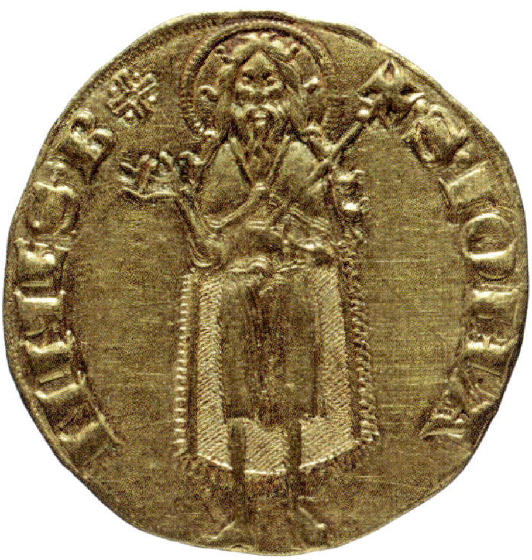

Florence, Biblioteca Medicea Laurenziana, Mediceo Palatino 219

Mexico, 16th century (1577)
fol. 335v, detail

> The value of cacao
> One turkey, 200 cacao beans.
> One canoe, 100 cacao beans.
> One rabbit, 30 cacao beans.
> One pumpkin, 4 cacao beans.
> One fish, 3 cacao beans.
> One avocado, 3 cacao beans.
> One egg, 3 cacao beans.
> One tomato, 1 cacao bean.
> One tamal, 1 cacao bean.
> The favours of a prostitute, 20–50 cacao beans.
> A slave, if he could sing and dance could be bought for 2400–4000 cacao beans, otherwise only between 600 and 1000 cacao beans.
> Philippe Nondedeo, 'El Chocolate entre fiestas y rituales mayas', *Artes de México*, no. 103, Mexico, 2011, pp. 35–45: 45.

The cacao tree has a delicate constitution, so it requires special conditions and considerable care for it grow and produce fruit. Cacao production was thus restricted by environmental factors. This meant that vast regions of Mesoamerica were outside the geographically suitable area for growing cacao and, consequently, it could not be produced on a large scale. This fact is of the utmost importance for understanding the economic role of cacao, as its fragility meant that it was a rare and priceless commodity.

Although cacao was used as currency for trading in markets and for other transactions in daily life, it was not the only medium of exchange. In addition to gold and precious feathers, which were considered to be valuable currency, the Mexica had a special type of cape made of cotton or henequen called *cuachtli*, which was also used for the payment of tribute in place of cacao. It is documented that only five of the thirty-eight provinces controlled by the Mexica paid tribute in cacao, most likely because they were regions that produced beans for local consumption. [R.V.]

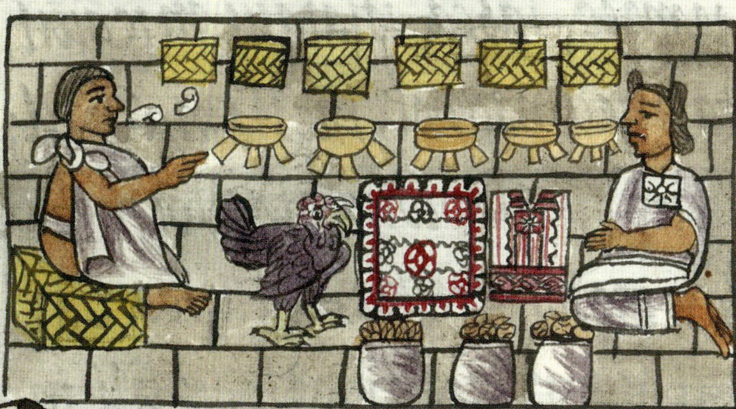

Despues, de auer comprado, y jũtado, todo lo necessario: luego da ua noticia deste banquete asus parientes, y amigos, viejos, y viejas para que le ayudassen con sus personas aseruir, el banquete: tambien daua noticia, y combidaua

POCHTECAS

Florence, Biblioteca Medicea Laurenziana, Mediceo Palatino 219

Mexico, 16th century (1577)
fol. 316r, detail

The *pochtecas* were the Mesoamerican couriers, known as the 'long-haul' merchants, who undertook long journeys to transport the valuable cacao beans and other goods. They played an indispensable part in the trade and procurement of rare and valuable goods for the Mesoamerican empire. The reason was very simple: since cacao was not naturally available in places like central Mexico, access to it was mainly through trade, conquests and the *pochtecas* themselves. The region of Xoconusco (Soconusco), conquered by the Mexica, was extremely valuable as it produced the most cacao in southern Mexico, but was over 800 kilometres from Tenochtitlán. The *pochtecas* were responsible for transporting the cacao tribute collected from the hot, tropical regions to the Valley of Mexico. The *carga* was the unit of measurement used for the cacao trade, equivalent to 24,000 beans, corresponding to a weight of about 20 kg. This was the maximum load the *pochtecas* could carry on their shoulders during their gruelling journeys. Over 40,000 *cargas* of cacao were carefully guarded in the 'Casa del Cacao' of Moctezuma II at the time of the arrival of the Spaniards. [R.V.]

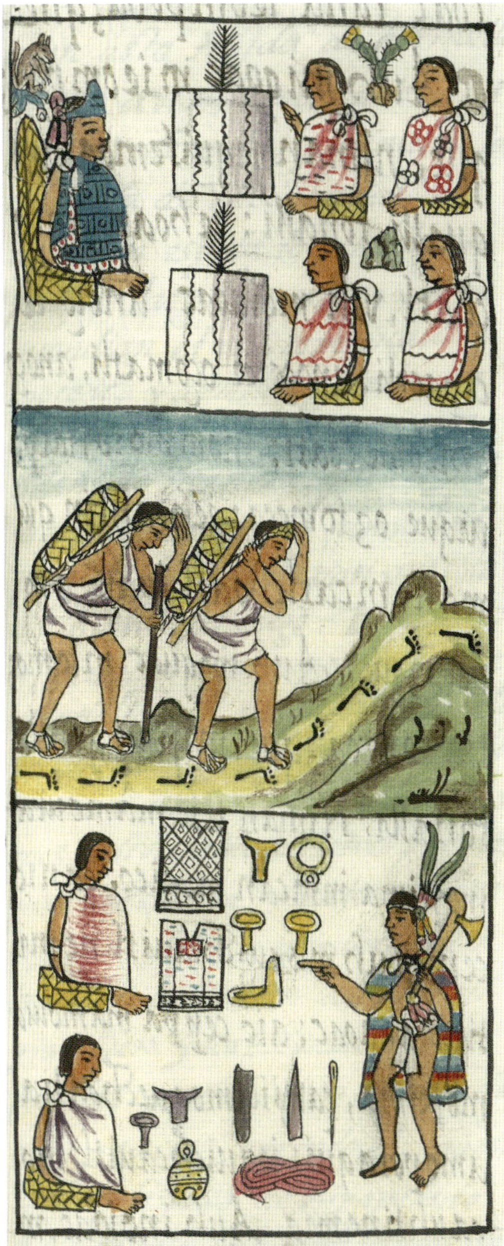

FALSAS

Florence, Biblioteca Medicea Laurenziana, Mediceo Palatino 220

Mexico, 16th century (1577)
fol. 48v, detail

El mal tratante véndelas falsas, porque las cuece y aun las tosta para que parezcan buenas, y a las veces échalas en el agua para que se paren gordas. E hácelas como cenicientas o pardas, que son las mejores almendras, para engañar. Tiene también este modo para adoballas: que las que son nuevas, para que parezcan gordas, suélelas tostar en la ceniza caliente, y después las envuelve con greda o con tierra húmeda, para que las que parecían menudas parezcan gordas y nuevas. Otro modo tiene para enganar: que en las cáscaras de las almendras mete una masa negra o cera negra que parece sea semejante al meollo dellas, y algunas veces los cuescos de aguacates los hacen pedazos, y redondéalos, y ansí redondeados los meten en las cáscaras vacías de las almendras.
('The evil cocoa merchant sells fake seeds. He cooks and roasts them to make them look good, and sometimes throws them in water to make them look fat. In order to deceive, he makes the seeds ashy or brown, which are the best ones. He also has this way of disguising them: he roasts those that are new to make them look fat, and then he wraps them in clay or damp earth, so that the small ones look fat and new. He also has another way to deceive: he puts a black mixture or wax in the seed shells that looks similar to their contents, and sometimes he breaks the avocado shells into pieces and then he puts them rounded off in the empty almond shells').
Bernardino de Sahagún, Mediceo Palatino 220, fols. 48v–49r.

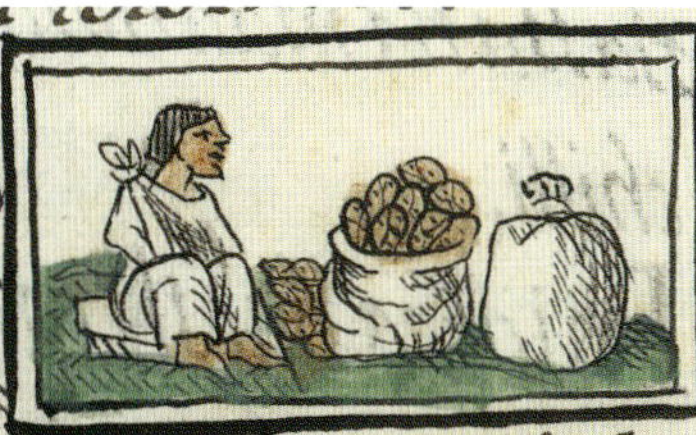

¶El maltratante vende las
falsas, porque las cueze, y aun
las tosta paraque parezcan
buenas. y alas vezes echalas
enel agua, paraque se paren
gordas. E hazelas como cene
zientas, o pardas, que son las
mejores almendras para en
gañar. Tiene tambien este
modo, para adouallas, que
las que son nueuas paraque
parezcan gordas, suele las tos
tar enla ceniza caliente, y des
pues las embuelue congreda
o contierra humeda, paraq̃
las que parecian menudas,
parezcan gordas, y nueuas:

CONVENTOS

Attributed to Johannes Vingboons and workshop, after Juan Gómez de Trasmonte, 1628

View of Mexico City

Amsterdam, *c.* 1640–68

Line-and-wash drawing on canvas-backed paper, 60.1 × 47.1 cm

Florence, Biblioteca Medicea Laurenziana, Carte di Castello, 52

On the site of the pre-Hispanic city, in 1523, the Spanish founded a new city, laid out by Alonso García Bravo (1490–1561). The map shows the basic layout divided into two parts: a central one, where the Spanish resided, and a peripheral one, where four indigenous quarters were located. The city had few streets and most of them were crossed by canals. The palaces of Moctezuma II were preserved, and the old Aztec square was adopted as the nerve centre of the city. The cathedral took almost three centuries to complete, rising above the old Aztec temple as an emblem of the subjugation of the indigenous religion and the glorification of the Catholic Church. The other Mexica temples served as foundations for numerous other churches and convents. In the city, there were 18 convents for religious males, 14 convents for religious females, 8 hospitals, 4 colleges and several parishes, in addition to the cathedral, the Viceregal Palace, the Cabildo house and the University facing onto the Plaza del Volador. The largest convents were dedicated to St Augustine, built by Augustinian friars between 1539 and 1560, characterised by its fortress-like appearance, with a Plateresque façade, and that of St Francis, founded in 1524 by the Franciscans, the first religious order to arrive in New Spain to evangelise the indigenous people who survived the conquest.

From 1523, Mexico City also became the main hub for the spread of Franciscan missions in the New World. [R.V.]

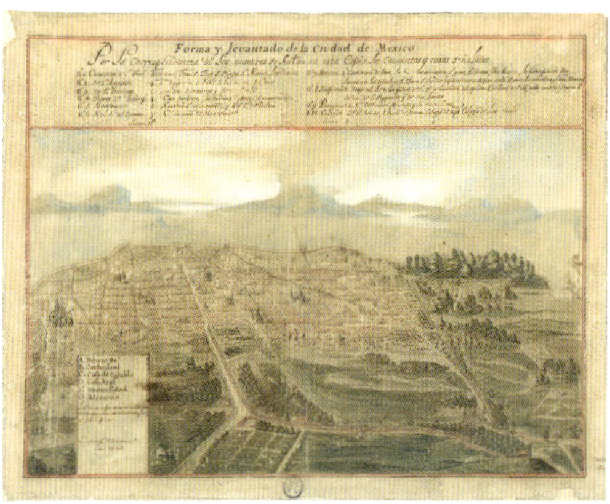

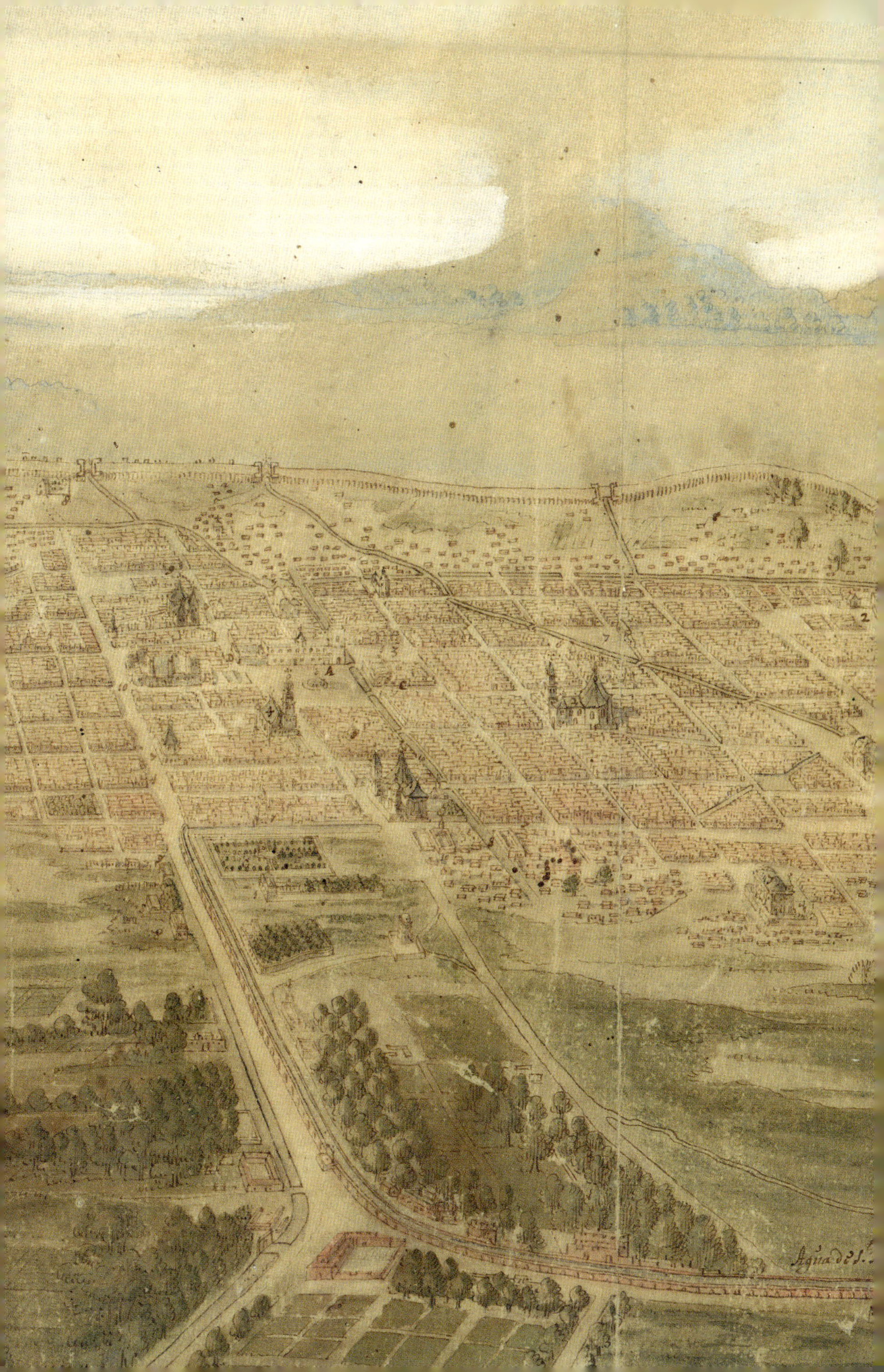
Agua de

Florence, Biblioteca Medicea Laurenziana, Mediceo Palatino 220

Mexico, 16th century (1577)
fol. 394r, whole and detail

The conquest had a twofold purpose: political and spiritual. Immediately after the discovery of the New World, the Borgia Pope, Alexander VI, granted dominion over the discovered lands in the Papal Bull *Inter Caetera* to King Ferdinand II of Aragon and Queen Isabella I of Castile, thereby lending the conquest a divine foundation. Religion emerged as the main justification for American colonisation; in order to establish Spanish rule in the conquered villages, taxes were levied and laws enacted so as to impose knowledge of the Christian faith on the natives, thus initiating the process of colonisation and land appropriation and inaugurating the viceroyalty of New Spain. The creation of the *encomiendas* was a very astute ploy to justify the system of economic exploitation in spiritual terms. The *encomendero* was appointed by the Crown to educate a group of people, civilise them and integrate them into the Catholic religion. His mission was benevolent, but in return the *encomendero* could avail himself of the physical toil of the people in his charge as compensation for the great service he was rendering them, that is, educating and catechising them. The weight of religion for the Aztecs was enormous, 'it was so great that we can say, without exaggeration, that their whole existence revolved around religion and there was not a single manifestation of public and private life that was not imbued with a religious feeling' (Alfonso Caso, *La religión de los aztecas*, Mexico 1936, p. 55).

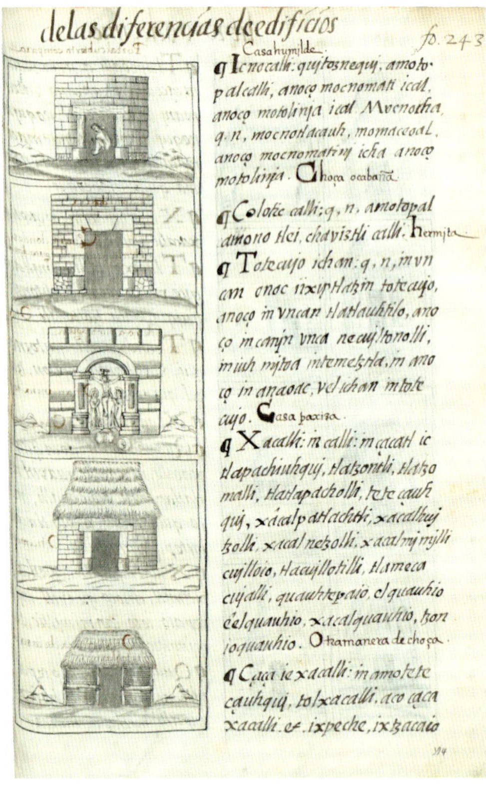

It is therefore through this prism that the history of the Spiritual

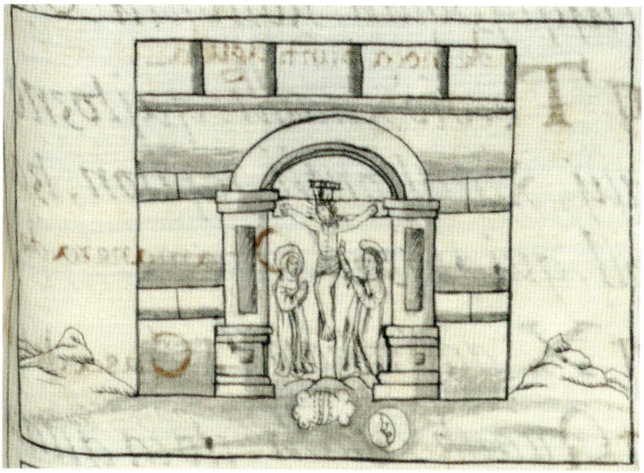

Conquest of the New World has to be viewed, implemented through the missionary expeditions of the mendicant orders. With the Papal Bull *Alias felicis* of 25 June 1521, Pope Leo X, the first of the Medici family to be Pope, granted the Franciscan Order official authorisation to launch the work of evangelisation in New Spain. A year later, with Pope Adrian VI's Papal Bull *Exponi nobis fecisti*, this permission was also extended to the other mendicant orders, including the Dominicans, Augustinians and Jesuits. In 1523 the first real evangelising mission was undertaken in the New World with the missionary expedition of three Franciscan friars: Johann Van den Auwera, Johann Dekkers and Peter of Ghent. They were the trailblazers of the Spiritual Conquest – this was the same year in which Cardinal Giulio de' Medici became Pope Clement VII. The three Franciscans, originally from the region of Flanders, left Spain together and arrived in Veracruz on 3 August 1523. From there they made their way to Mexico City, where they were welcomed by Hernán Cortés. The friars devoted themselves zealously to the conversion of the indigenous peoples, using the Nahuatl language and a special pictographic writing system. They founded convents, churches and schools in Texcoco and Tlatelolco, where they also taught arts and crafts to the natives. With the conviction that the best means of converting people to Catholicism was through the natives themselves, they concentrated their work on young people and children. To spread the religion, they employed various methods such as drawing, singing, theatre, music, setting a good example, poverty and work.

A year later, at the express wish of Cortés, another twelve Franciscan friars, later known as 'the Twelve Apostles of the Spiritual Conquest of New Spain', disembarked in Veracruz on 13 May 1524. In the same year or, more probably, at the beginning of 1525 (the documented dates are not very precise), these twelve friars had discussions with some wise Mexicans who had survived the conquest, in memorable talks during which they laid the foundations for the new evangelisation, in order to reveal the 'God for whom one lives'. Indeed, seven years later, according to Catholic tradition, on 12 December 1531, on the hill of Tepeyac, today north of the Mexican capital, the iconic image of Our Lady of Guadalupe appeared to the indigenous Juan Diego Cuauhtlatoatzin, who had just been baptised into the Christian faith by the Franciscans.

In the mid-16th century, these 'coloquios de los doce' (Discussions of the Twelve) were documented and collected by the founder of modern anthropology, who for forty-five years (1540–85) wrote about the Mexica in Texcoco, Tlatelolco and Mexico City: the Franciscan friar Benardino de Sahagún. [R.V.]

4. NATURALIA

Book X, Vices and virtues, illnesses and cures
'de la general historia, en que se trata de los vicios, y virtudes, ansi espirituales, como corporales, de toda manera de personas'

Book XI, Animals, plants, minerals
'de las propiedades de los animales, aves, peces, árboles, yervas, flores, metals, y piedras, y de los colores'

BERNARDINO DE SAHAGÚN

> With the Spanish Conquest, the indigenous world was plunged into silence. With its history and culture extinguished, its religion dead, it now embarked on a long process of acculturation, as part of a new faith that the victors attempted to spread, not only by preaching but by force. However, although defeated in material terms, the pre-Columbian world ended up surviving covertly, penetrating deep into much of the new artistic expression, from colonial times to the present day. In the rescue and dissemination of the literary creations coming from regions ranging from Mexico to the Inca Empire, credit must be given, as mentioned, to religious men, the Franciscans above all, keen scholars of indigenous cultures, and to the early chroniclers, no small number of whom celebrated the splendour of the indigenous production.
> Giuseppe Bellini, *Tra Medioevo e Rinascimento. La poesia nell'America conquistata*, Salerno, 2003, p. 146.

At the dawn of the Renaissance, Bernardino de Ribeira (1499–1590) was born in the small town of Sahagún de Campos, Spain. After studying at the University of Salamanca, he decided to join the Franciscan Order and embarked for Mexico in 1529. Here he was to stay for good, giving Latin lessons, writing and exploring the Nahuatl language. Wishing to base the evangelisation effort on the foundation of knowledge of 'the Other', around 1540 he began to collect and compile information on the Mexica. The missionary's intention was to act like a doctor who tries to identify the origin of his patient's illness. First it was in the village of Tepepulco, then in Tlatelolco and finally in México-Tenochtitlán, where he conducted his research. With the invaluable support of his disciples, the grammarians (*tlacuilos*), indigenous and trilingual scribes of Latin, Spanish and Nahuatl, of the College of Santa Cruz Tlatelolco, Sahagún compiled a vast body of information about 'the natural, human and divine things' of the ancient Mexicans.

In order to gather the information, he required, Sahagún developed a research method that is considered the forerunner, if not the originator, of cultural anthropological research as it is practised today. He started with questionnaires addressed to elderly sages – the *principales* – who had survived the Spanish conquest, through conversations and interviews. Their oral answers were translated into pictograms – as this was their mode of writing before the conquest – and later interpreted, expanded and transcribed into alphabetical Nahuatl, or Latin characters in the language of the Mexica.

By means of this process, documents made up of drafts, subsequent revisions and editing were produced, known as *Primeros Memoriales* (1558–9, preserved in the Códices Matritenses), *Manuscrito de Tlatelolco* (consisting of *Primer manuscrito de Tlatelolco*, 1561–2; *Memoriales en tres columnas*, 1563–5; *Memoriales con escolios*, 1565). His linguistic expertise is evident in the lists of Nahuatl words accompanied by explanations and there are also numerous annotations in the margins made by Sahagún. His field research continued for several years, from 1558 until almost 1570.

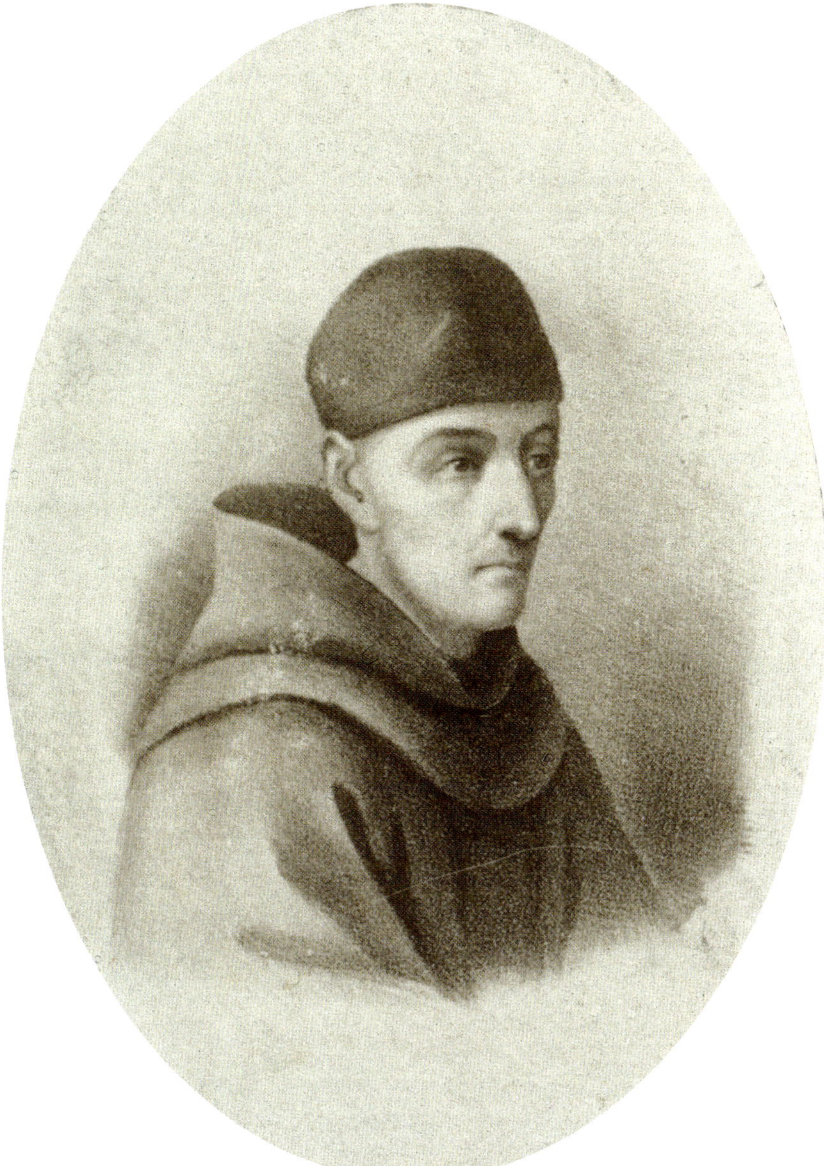

Friar Bernardino de Sahagún.

In 1569, Sahagún began to arrange this information into twelve books and also worked on other texts including *Manuscrito* (1569), *Sumario* (1570) and *Breve compendio de los ritos idolátricos de Nueva España* (1570). Subsequently, the friar was obliged to hand these documents over to the Franciscan authorities for revision and it was only in 1575 that they were returned to him, thanks to the intervention of Friar Rodrigo de Sequera, who had just arrived in Mexico as the new Commissary General of the Franciscan Order. Sequera provided Sahagún with the necessary support to complete the work, stating clearly that it should be translated into Spanish.

During this period, the writing that was produced saw the abandonment of the traditional pre-Hispanic format (*amoxtli*) – like the Dresden Codex, which can be folded accordion-style and is made from the traditional indigenous paper *amatl* bleached with lime – and took on the characteristics of European books, both in terms of format and the use of paper, which was of European origin. The tools and materials used in manuscript production also included brushes, pencils and quills, along with pre-Hispanic and European pigments of mineral, vegetable and animal origin.

They enriched and enhanced the bilingual narrative with a prodigious amount of extraordinarily effective illustrations. These are interpretations or readings of what is shown in Nahuatl, by indigenous painters trained in European art. A perfect dialectic *entre dos mundos* (between two worlds).

> The process of collating the information to create this indigenous encyclopaedia took almost thirty years, but the process of completing this work was a race against death.
> Diana Magaloni Kerpel, 'Painters of the New World: The Process of Making the "Florentine Codex"', in *Colors between Two Worlds. The Florentine Codex of Bernardino de Sahagún*, proceedings of the conference (Florence, 12–13 June 2008), Milan, 2011, pp. 47–76: 49.

Sahagún's work – a choral work – continued for some thirty years, even when the plague raged around the *scriptorium*: in 1576, during the translation of the work into Spanish, the city was struck by an epidemic of biblical proportions, causing the death of more than 80% of the indigenous population of the Valley of Mexico. Sahagún and the surviving students of the Colegio de la Santa Cruz determined to continue the work. However, due to the persistence of the plague, the world they were depicting was gradually losing its vibrant colours, turning into a black and white palette in its final stages.

Completed in 1577, the work was given the title *Historia general de las cosas de Nueva España.*

Today, almost 500 years later, it is recognised as the most influential manuscript of the early colonial period in Mexico and the most reliable source of information on the Mexican culture, the Aztec empire and the conquest of Mexico. On account of the method he devised for his research and the vast amount of evidence he collected on the culture of the Mexica peoples, Bernardino de Sahagún is acknowledged as the father of modern anthropology. [R.V.]

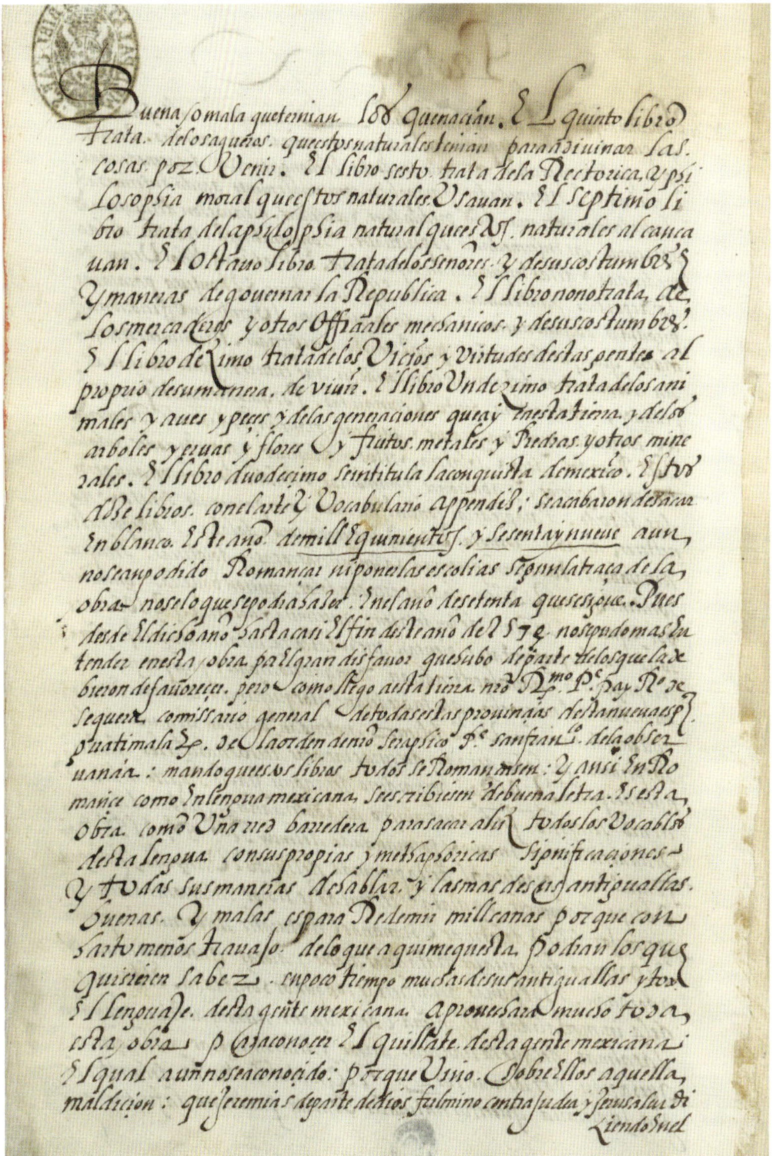
Buena, o mala que tenian los que nacian. El quinto libro trata delos agueros que estos naturales tenian para adiuinar las cosas por venir. El libro sesto trata dela Rectorica y philosophia moral que estos naturales usauan. El septimo libro trata dela phylosophia natural que estos naturales alcançauan. El octauo libro trata delos señores y de sus costumbres y maneras de gouernar la Republica. El libro nono trata delos mercaderes y otros officiales mecanicos y de sus costumbres. El libro decimo trata delos vicios y virtudes destas gentes al proprio de su manera de viuir. El libro undezimo trata delos animales y aues y peces y delas generaciones que ay en esta tierra y delos arboles y yeruas y flores y frutos, metales y piedras y otros minerales. El libro duodecimo se intitula la conquista de mexico. Estos doze libros con el arte y vocabulario appendiz se acabaron de sacar en blanco. Este año de mill quinientos y sesenta y nueue aun no se an podido romançar ni poner las escolias segun la traça dela obra: no se lo que se podra saber del año de setenta que se sigue. Pues desde el dicho año hasta casi el fin deste año de 1575 no se pudo mas entender en esta obra por el gran disfauor que huuo de parte delos que la deuieron de fauorecer. pero como llego a esta tierra nro Rmo P. fray Ro de Sequera comissario general de todas estas prouincias desta nueua españa y guatimala &c. de la orden de nro Seraphico P. sanfran.co dela obseruancia: mando que estos libros todos se romançassen: y ansi en Romance como en lengua mexicana se escribiesen de buena letra. Es esta obra como una red barredera para sacar a luz todos los vocablos desta lengua con sus propias y metaphoricas significaciones y todas sus maneras de hablar, y las mas de sus antiguallas buenas y malas es para redemir mill canas porque con harto menos trabajo delo que aqui me questa podran los que quisieren saber en poco tiempo muchas de sus antiguallas y todo el lenguaje desta gente mexicana. aprouechara mucho toda esta obra para conocer el quilate desta gente mexicana el qual aun no se a conocido: porque vino sobre ellos aquella maldicion: que Jeremias de parte de dios fulmino contra Judea y Jerusalen diziendo: el

'… to know the value of these Mexican people'. Florence, Biblioteca Medicea Laurenziana, Mediceo Palatino 218, Mexico, 16th century (1577), fol. 1v.

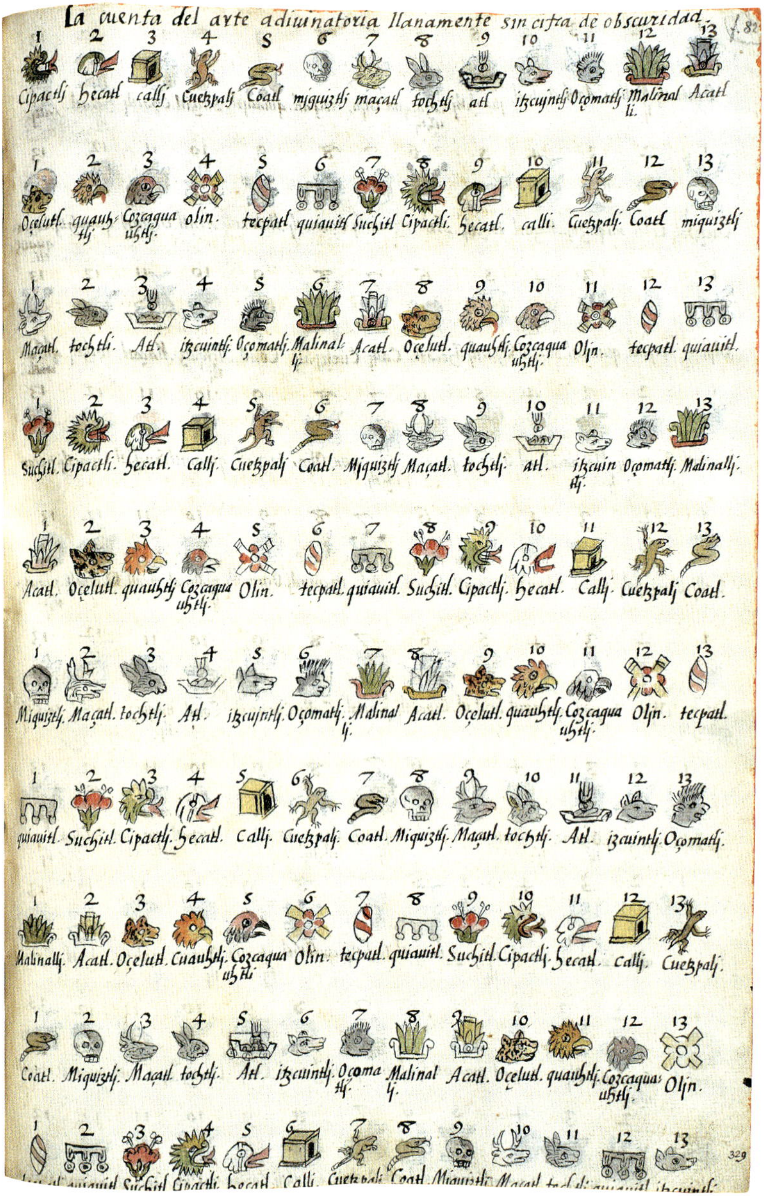

Florence, Biblioteca Medicea Laurenziana, Mediceo Palatino 218, Mexico, 16th century (1577), fol. 329r.

Florence, Biblioteca Medicea Laurenziana, Mediceo Palatino 220

Mexico, 16th century (1577)
fol. 155v, detail of the *ocelotl* (or American jaguar)

The questionnaire given by Friar Bernardino to his researchers about animals (Book XI) is particularly significant. Here are the questions:

> 1. The name or names of the animal.
> 2. Which other animals does it resemble?
> 3. Where does it live?
> 4. Why does it receive this name?
> 5. What does it look like?
> 6. What habits does it have?
> 7. What does it feed on?
> 8. How does it hunt?
> 9. What sounds/noises does it make?
>
> Alfredo López Austin, 'Estudio acerca del método de investigación de fray Bernardino de Sahagún', *Estudios de Cultura Náhuatl*, no. 42, 2011, pp. 353–400: 385. [R.V.]

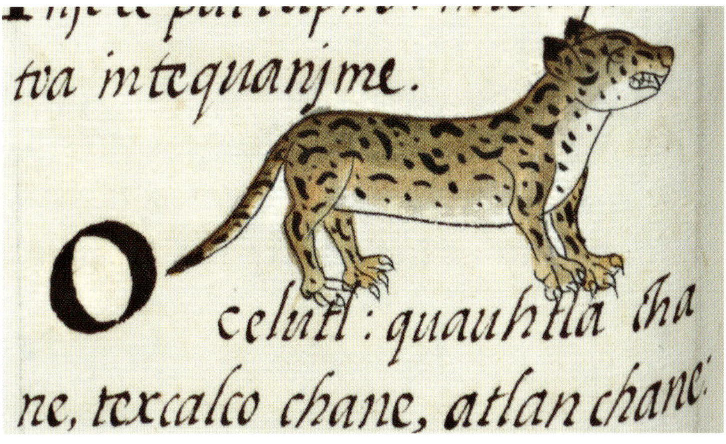

Florence, Biblioteca Medicea Laurenziana, Mediceo Palatino 220

Mexico, 16th century (1577)
fol. 43r, detail

Organised in twelve books and divided into three volumes – initially conceived as four tomes – this monumental work was structured like an encyclopaedia. Divided by subject, the Codex provides detailed descriptions, ranging from the theological-religious cosmovision and rituals, including human sacrifices aimed at establishing a bond with the deities, to the practices and customs of a people steeped in a rich tradition. It also describes the rich Mexican natural world and concludes with the story of the conquest, told from the point of view of the vanquished.

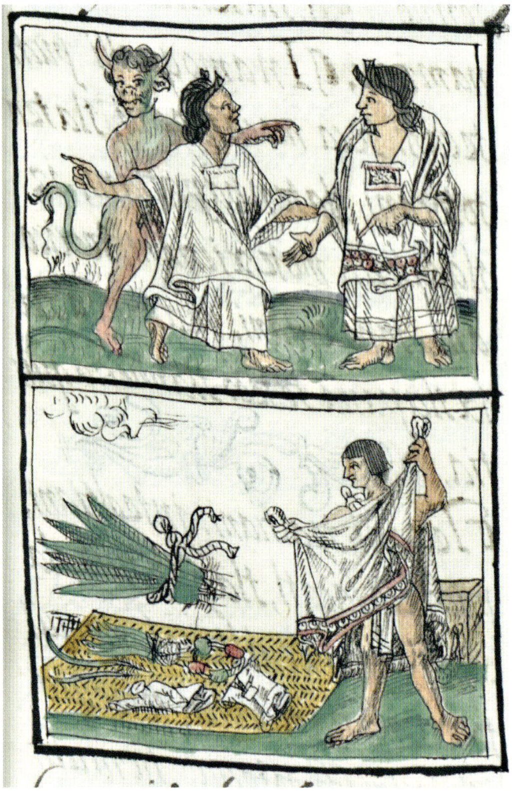

'… it is a carefully and skilfully crafted work of art whose material qualities and historic context bear the indelible marks of an epic battle for survival' (Diana Magaloni, in *Il Mondo degli Aztechi nel Codice Fiorentino*, Florence, 2007, p. 12).

The contents of the book are laid out in two parallel columns: in the left column, the text is in Spanish, while in the right column it is in Nahuatl.

Books I, II and III deal with the religious sphere of the natives, deities, myths, festivals, sacrifices and idolatry:

I. 'que se trata de los dioses que adoraban los naturales desta tierra que es la Nueva España'.

II. 'que trata del calendario, fiestas y cerimonias, sacrificios y solenidades que estos naturales désta Nueva España hacían a honra de sus diose'.

III. 'del principio que tuvieron los dioses'.

Books IV and V deal with astrology and divination:

IV. 'de la astrología judiciaria o arte adivinatoria indiana'.

V. 'que trata de los agüeros y prenósticos que estos naturales tomaban de algunas aves, animales y sabandixas para adivinar las cosas futuras'.

Book VI collects prayers and solemn utterances that illustrate types of Nahua rhetoric:

VI. 'de la retórica y filosofía moral y teología de la gente mexicana, donde hay cosas muy curiosas tocantes a los primores de su lengua y cosas muy delicadas tocantes a las virtudes morale'.

Book VII regards the Sun, Moon and stars:

VII. 'que trata de la astrología y filosofía natural que alcanzaron estos naturales de esta Nueva España'.

Book VIII is devoted to the history of noblemen and rulers:

VIII. 'de los reyes y señores y de la manera que tenían en sus electiones y en el gobierno de sus reinos'.

Book IX deals with trade, merchants and artisans dedicated to goldsmithing and other crafts:

IX. 'de los mercaderes, oficiales de oro y piedras preciosas y pluma rica'.

Book X describes Mexican society, vices and virtues of the people, diseases and remedies:

X. 'de la general historia, en que se trata de los vicios, y virtudes, ansi espirituales, como corporales, de toda manera de personas'.

Book XI, with the greatest number of pages, is a treatise on natural history:

XI. 'de las propiedades de los animales, aves, peces, árboles, yervas, flores, metals, y piedras, y de los colores'.

Book XII, describes the Spanish conquest and the fall of Tenochtitlán from the perspective of the Mexica:

XII. 'de la conquista de la Nueva España, que es la ciudad de México'.

The approximately 2,500 illustrations that appear next to the alphabetical texts in Spanish and Nahuatl, done in coloured ink within small black frames and placed in the left-hand column, are the 'third narrative of the Codex' (Kim Richter, 'Memoria del Nuevo Mundo', Conference, Florence 2023).

One word is omnipresent in 11 of Sahagún's 12 books: cacao. [R.V.]

Florence, Biblioteca Medicea Laurenziana, Mediceo Palatino 220

Mexico, 16th century (1577)
fol. 275, detail

'Cacao' is a word that is continuously repeated in Sahagún books. The origin of the word is probably 'kakawa', datable to 4,000 years ago, from the languages of the Mixe-Zoques, a culture derived from Olmec, which developed in the current region of the southern state of Veracruz, north-west of Tabasco and the Isthmus of Tehuantepec. From here, it spread to the Maya area ('kakaw') and later extended to all Mesoamerican languages, including Nahuatl ('cacao'), the language of the Mexica people.

It refers to the cacao tree, its fruit and the drink derived from the cacao beans, which has deep symbolic, ritual and therapeutic connotations.

The attribution of the name by the Olmecs is considered the first linguistic evidence of the existence of cacao. The richness of the archaeological, linguistic and iconographic findings, which attest to the consumption of cocoa as a beverage 3800 years ago, strengthens the theory that its first real domestication took place in Mesoamerica and that it was also the Olmecs who developed the process of chocolate production. This theory was put forward by archaeologist Michael D. Coe and historian Sophie D. Coe in *The True History of Chocolate* (1996).

The latest genomic research has revealed a new archaeological culture called Mayo-Chinchipe, where cocoa consumption can be traced back to 3500 BC, at the site of Santa Ana (La Florida), in the forests of the upper Amazon. However, no evidence has emerged as to the existence of a specific name for the cacao tree in that region. Thus, although there is evidence to establish the point of origin of a cacao variety in the Amazon, it is not possible to prove that this area was also the place of its initial domestication. The act of naming the item in question in the domestication process is closely linked to the socio-cultural dynamics of human groups: words are generated to name things or define concepts. It is more likely that they merely consumed the white pulp enveloping the seeds of the fruit and enjoyed the seeds alone, without processing the cacao as such.

Pending further archaeological research that may reveal other historical scenarios in the future, the fact stands that it was the Olmec civilisation that pioneered the cacao culture. It later became a commodity of religious, social, medicinal and economic importance throughout Mesoamerica. [R.V.]

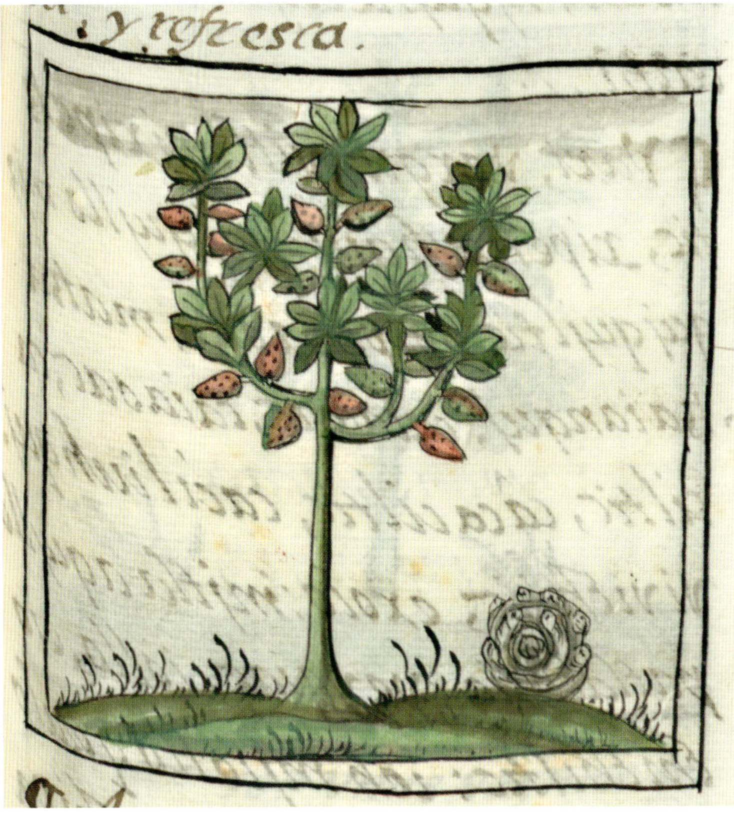
y refresca.

KAKA(W)

Florence, Biblioteca Medicea Laurenziana, Mediceo Palatino 220

Mexico, 16th century (1577)
fol. 214v

With the Maya, cacao became the subject of many artistic items and artifacts. All manner of containers existed to preserve it, linked to ceremonies of daily life. The first written evidence of the use of the word cacao is attested to in the Classic Maya culture (which had learned about cacao from the Olmecs themselves, around 1000 BC). The glyph commonly used to write *kakaw(a)* is composed of three phonetic parts in which, according to Maya custom, the last vowel is silent.

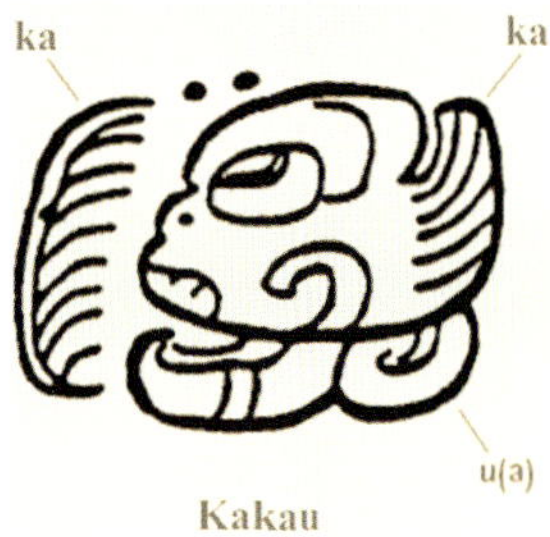

Cacao glyph.

Despite numerous studies on the subject, the cacao glyph still harbours many mysteries. In the numerous archaeological finds of cylindrical pottery that accompanied Maya nobles into the afterlife, a fish with two fins repeatedly appears. David Stuart, in *Chocolate in Mesoamerica*, offers one of the first technical explanations: the syllable 'ka', represented by a fin, takes its image from the glyph 'kay', fish, hence the writing of the word with the pictogram of a fish; thus the two fins are the repetition of the syllable 'ka', in twin syllables: 'ka- ka-'; the last syllable of the glyph, 'wa (ua)', represented with a sort of shoot or sprout, takes on the meaning of 'renewal', completing the glyph 'kakau(a)'.

The cacao 'fish' is directly linked to the tradition passed down in the *Popol Vuh*, the Maya-Quiché book of genesis, which recounts that the first cacao tree flourished when the heroic twins of the Ahpú dynasty, Hunahpù and Xbalanqué, rose from the primordial waters of the Underworld after passing through a stage as fish. These heroes of the *Popol Vuh* were presumably also the inventors of the processes needed to turn the fruits of the cacao tree into chocolate. Thus, the water twins lent their fins and wavy shape to the cacao glyph, by means of which the drink was elevated to the status of an elixir of resurrection.

For this reason, it can be asserted that the cacao tree grows, both in a real and a symbolic sense, in the shadows, concealing ancestral notions of death and resurrection. [L.H.–R.V.]

llama totomjchi, que qujere dezir ave pez: dize se aue: porque tiene la cabeça como aue, y el pico como aue, y pica como aue: y tiene las alas largas como pez, y la cola como pez

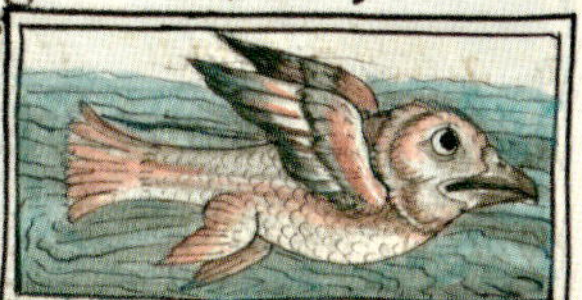

¶ Ay vn pez en la mar que se llama vitzitzilmjchi: llamase ansi, porque tiene el piqujllo muy delgado, como el auecilla, que se llama zinzon, que anda chupando las flores.

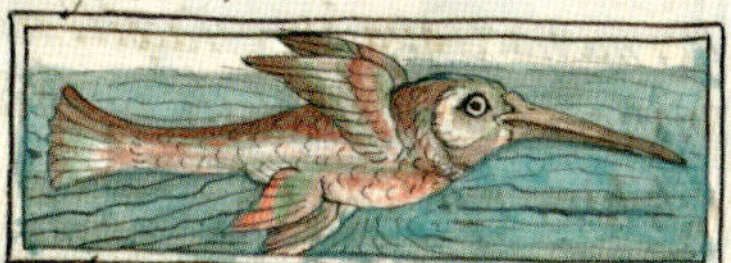

¶ Ay otro pez en la mar que se llama papalomjchi, que qujere dezir pez como mariposa, porque es de la hechura de mariposa.

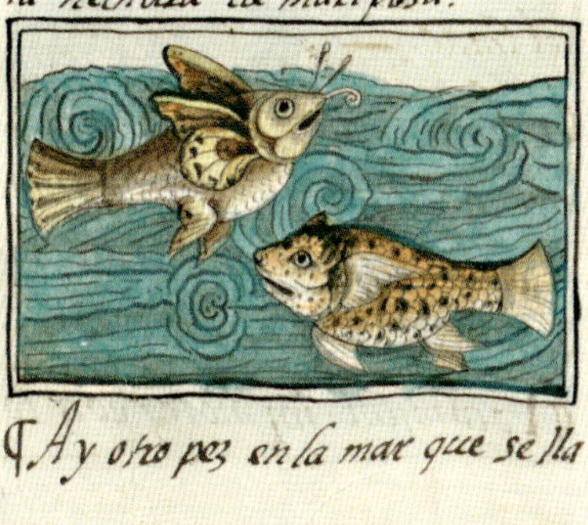

¶ Ay otro pez en la mar que se lla

tomjchin. vel mjchi, achi a
tlapal viac. Injc mjtoa toto
mjchi: injtzontecon vel iuh
qujn tototl itzontecon icac: ve
tenvitztic, temmjmjltic, tla
ponjanj, tlachoponja, tlape
tulon.

¶ Vitzitzilmjchi: vel mjchi, m
mjtoa vitzitzilmjchi. achi v
ac injatlapal, mjpatlanj
mjtlanelooaia. Auh mjten
vel vitztic, viac, vel iuhqu
vitzitziltentli mjten: tenvit
ponatic, tenvitzmalotic, ten
vitzmallotl

¶ Papalomjchi: veiapan tl
ne, caçan ie mjchi: auh inj
ca, mjc tlachioalli: vel iuh
papalotl, patlachtic, iacats
ne: injuhquj papalotl, am
tlapale; injatlapal vel iuhq
papalotl yiatlapal: in atl y
itic ienemj, iuhqujn patlan
çolonj.

¶ Ocelomjchi: no veiapã

CACAOAQUAUTIL

> Al árbol donde se hace el cacao llaman cacahuacuáhuitl. Tiene las hojas anchas y es acopado. Es mediano. El fruto que hace es como mazorcas de maíz, o poco mayores, y tienen de dentro los granos de cacao; de fuera es morado, y de dentro encarnado o bermejo. Cuando es nuevo, si se bebe mucho enborracha. Y si se bebe templadamente refrigera y refresca. ('The tree from which cacao is produced is called *cacaoaquauitl*: it has broad, cup-shaped leaves and produces medium-sized fruit, similar to corn cobs or slightly larger, containing the cacao beans. On the outside it is purple and on the inside it is flesh-coloured or vermilion. When young, if consumed in large quantities, it causes drunkenness; when consumed moderately, it has a cooling and refreshing effect').
> Bernardino de Sahagún, Mediceo Palatino 220, fol. 275r.

The Nahuatl term for the cacao tree is *cacaoaquauitl*. It is an evergreen that grows naturally in the tropical rainforest biome, in a belt stretching 20° north and 20° south of the Equator, with temperatures between 18 and 32 °C and precipitation between 1,500 and 2,000 mm per year. It grows protected in the shade of larger trees that shelter it from the sun and wind. It is a relatively small tree that reaches an average height of 6 metres. Depending on its size, there are four different names for it in the Nahuatl language: *cuauhcacahuatl*, the largest; *mecacahuatl*, *xochicacahuatl* and *tlalcacahuatl*, the smallest. At the time of the Mexica empire and the conquest, cultivation was limited to the specific geographical regions of its habitat, near the sumptuous aristocratic palaces and ceremonial centres of the cacao-producing cities.

Squirrels and monkeys (mainly the genus Ateles, *mono araña*) steal the pods to enjoy the pleasant-tasting white pulp which envelops the kernels, but avoid the beans themselves, thus contributing to their dispersal. 'The monkeys cannot be seeking the beans, which are made bitter by alkaloids, but the delicious pulp, which is probably what attracted humans to *T. cacao* in the first place' (Sophie D. Coe, Michael D. Coe, *The True History of Chocolate*, London, 1996, p. 22).

This tree belongs to the genus Theobroma, of the family Malvaceae. It has 22 species, of which only two are important for cultivation: *Theobroma cacao* and *Theobroma bicolor*. The species *Theobroma bicolor*, known as the 'Macambo tree', is used as a domestic plant and produces *pataxte*. The species *Theobroma cacao* produces the cacao fruit. Its initial wild distribution encompassed the vast regions of Mesoamerica and the northern and western portions of the Amazon basin, where it underwent an evolutionary process that led it to diverge into two distinct species:

1. *Theobroma cacao* from Amazonia. It produces hard, round fruits like small melons. It belongs to the 'forastero' varietal type and is a strong tree, highly resistant to disease and with a high yield of fruit. Its seeds are dark, very bitter and not very aromatic.

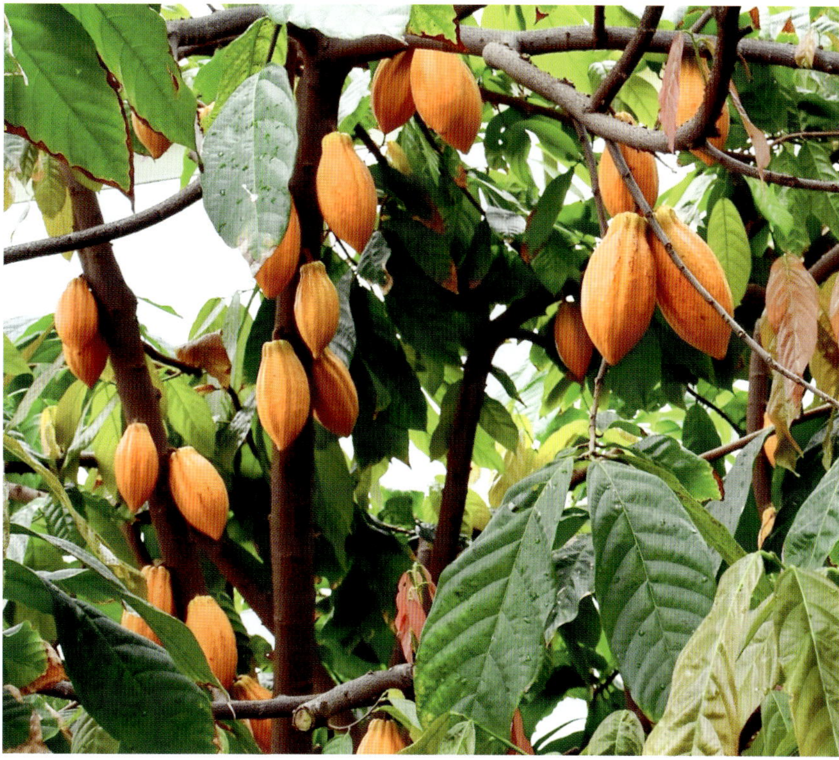

Cacao tree.

2. *Theobroma cacao* from Mesoamerica. It is distinguished by its long, pointed, wrinkled fruits with deep grooves. It belongs to the 'criollo' varietal type and is a delicate tree susceptible to disease, producing few pods, but its light-coloured seeds are very fragrant and not too bitter, with a very high-quality flavour and aroma. It was the species for which the rulers and nobles of ancient Mesoamerica had a particular predilection.

In the 18th century, the first *Theobroma cacao* hybrid was created on the island of Trinidad off the coast of Venezuela. A hurricane or disease caused the death of numerous trees of the 'criollo' varietal type previously planted in the area.

During the subsequent planting of 'forastero' trees to replace them, cross-pollination took place with the surviving 'criollo' trees, thereby giving rise to 'trinitario' cacao. [R.V.]

FLOR DE CACAO

As is often the case with tropical fruit trees, the cacao tree flowers mostly cauliflorously, that is, directly from pads on the trunk. The flowers, which are small and consist of five petals, are pollinated exclusively by tiny flies that thrive in this environment. Cacao flowers are rich in phenylethylamine (a chemical that naturally alleviates stress by stimulating the production of endorphins, responsible for the common feeling of well-being one gets when eating chocolate). Their perfume is also widely used industrially, due to the natural fragrance – described as a mixture of vanilla and tropical flowers – that they release at the moment of flowering.

It is estimated that only between 1% and 3% of the flowers subsequently develop into the fruit, which is commonly called 'cabossa' and often ignored because of its much more prized seeds. [L.H.]

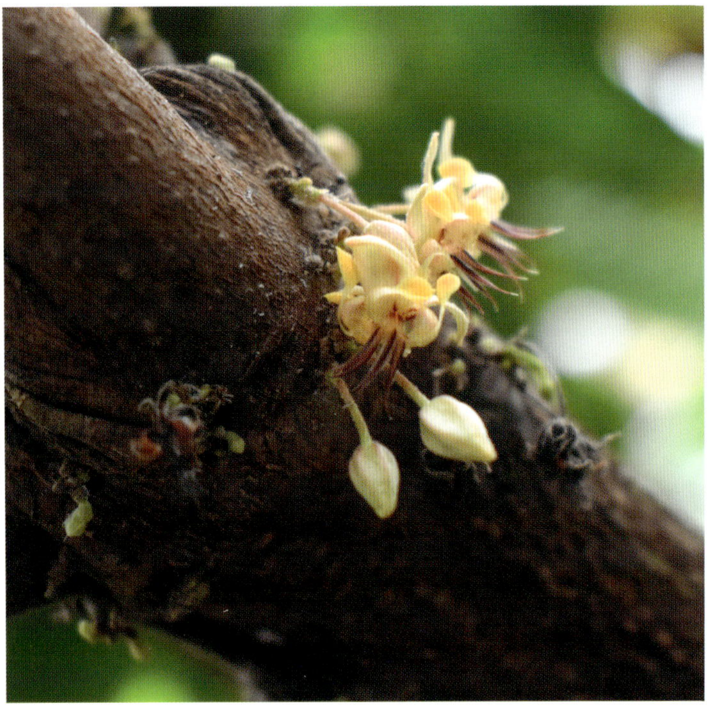

Flowers of the cacao tree.

FRUIT

Once the flower has been pollinated, the fruit develops: an elongated, ribbed pod called 'cabossa' or 'cabossis', which can measure up to 15–30 cm long by 8–10 cm wide. It grows directly on the trunk and its colour varies from yellow-green to red-brown, depending on the stage of ripeness. Each *cabossa* contains 25 to 40 almond-shaped seeds, called cacao beans, which are embedded in a sweet, gelatinous white pulp called mucilage.

In plantations, men have to pick and open the fruit to obtain the cacao beans. It is essential that harvesting is done with great care and attention, so as not to damage the pads that produce flowers and thus fruit. The harvest takes place twice a year, usually between November and January and between May and July. [L.H.]

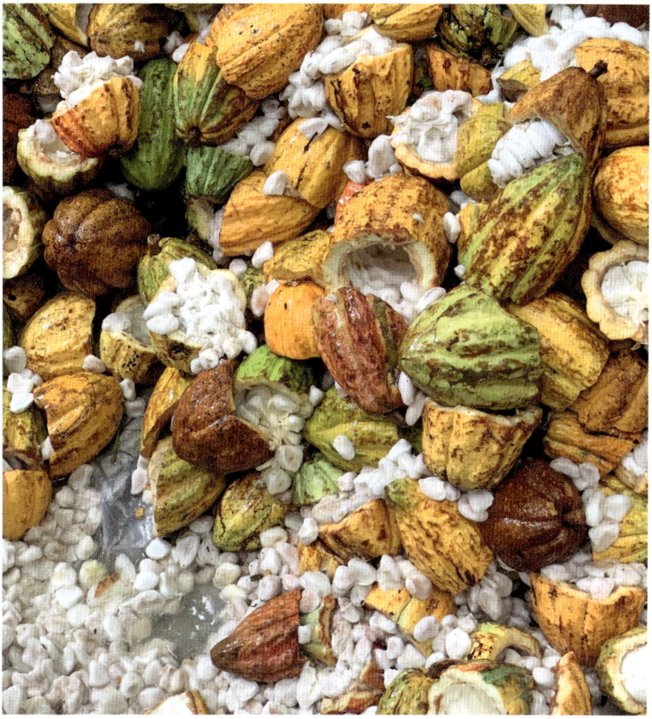

Cacao fruit, the *cabossas*.

CACAO

Once the cacao pods have been opened and the fruit extracted, there are four main steps in cacao processing: fermentation, drying, roasting and decortication.

> No matter what the level of technology, this sequence has been in force for at least three millennia, and still is followed in the modern world.
> Sophie D. Coe, Michael D. Coe, *The True History of Chocolate*, London, 1996, p. 23.

The fruit removed from the *cabossa* consists of cacao beans and the pulp that envelops them. This is then gathered, placed in piles directly in the harvest area or, in certain circumstances, in stacked wooden crates. This starts the spontaneous fermentation process, which can last from three to five days. From the first day, chemical and biological processes are activated. The pulp amalgamates with the seeds and becomes liquid, then there is a short germination phase followed by the rapid death of the seeds, caused by high temperatures and increased acidity. This phase gives the beans their characteristic chocolate flavour. Around the third day, the seed mass, which has to be turned from time to time, is between 45 and 50 °C and must remain at this high temperature for a few days in order to facilitate the absorption of the flavour acquired during germination.

Once fermentation is complete, the beans are dried traditionally on mats or trays left out in the sun, which takes one to two weeks, depending on the weather conditions. The enzymatic action triggered during fermentation is not interrupted during the drying process, during which the cacao beans lose more than half their weight.

Since the cacao beans are always of different shapes and sizes, it is important to select them so that they have a similar size within each roasting batch in order to ensure uniformity during the process. Roasting, which lasts between 70 and 115 minutes, takes place at temperatures ranging from 99 to 104 °C. During this phase, the beans take on a more intense shade of brown, become crumblier and lose their astringency. This is an essential step for the development of the required flavour and desired aroma.

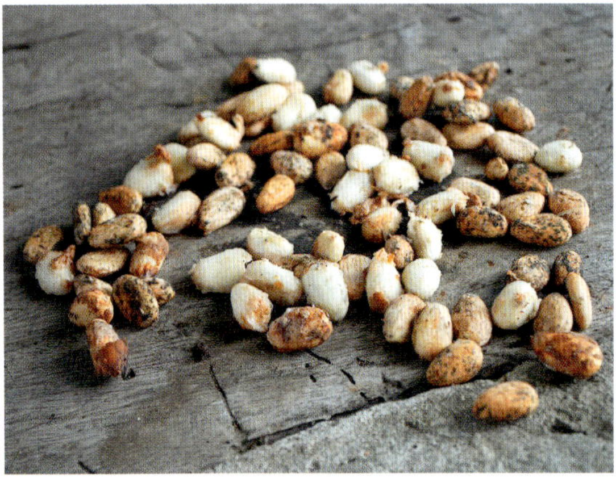

Cacao beans extracted from the *cabossa*.

The final step in obtaining processed cacao is to remove the husks that envelop the beans. Once this process is complete, the cacao is ready to be processed to produce chocolate and its myriad variations. [R.V.]

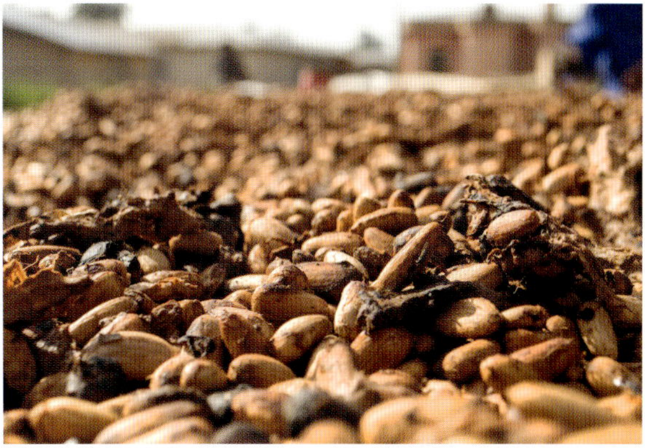

Cacao seeds undergoing fermentation.

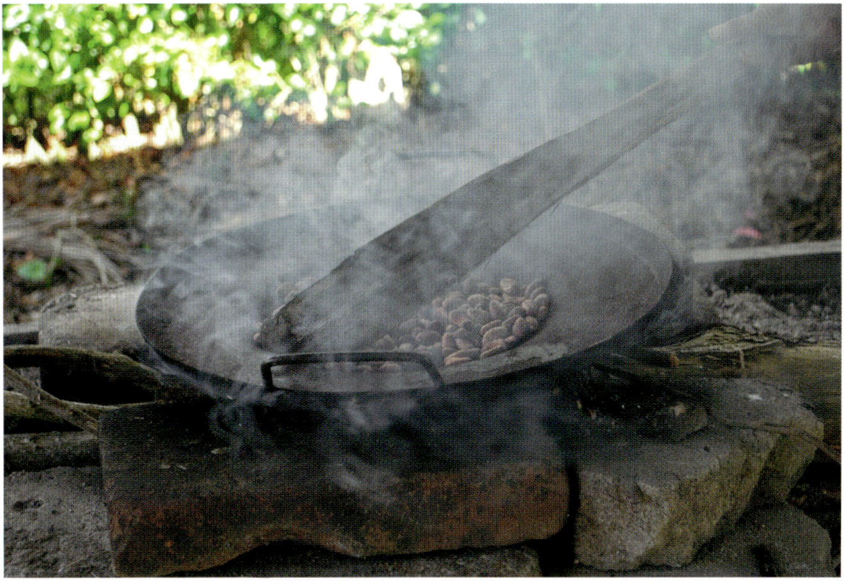

The roasting of the cacao beans.

Florence, Biblioteca Medicea Laurenziana, Mediceo Palatino 220

Mexico, 16th century (1577)
fol. 71v

Prior to the Conquest, chocolate was used in the preparation of a wide and complex range of beverages and foodstuffs, both solid and powdered, to which a wide variety of flavourings could be added.

In the Nahuatl language, the divine food was called *cacaoatl*, a term that translates literally as 'cacao water'. The Mexica consumed the chocolate cold, but the Maya and the Spanish colonisers preferred it hot. According to native sources in Sahagún, high-quality chocolate was called *atlaquetzalli* (precious water).

In Book X of Sahagún, where the accounts of sellers of various products in the market are recorded, in folio 71, the original recipe for chocolate is transcribed for the first time and from reliable sources. Sahagún placed this text next to ones about sellers of *atole*, a corn drink, *tequesquite*, a type of salt and, curiously, also near one about 'she who sells herself', the *ahuiani*:

> La que vende cacao hecho para beber muélelo primero en este modo, que la primera vez quiebra o machuca las almendras; la segunda vez van un poco más molidas; la tercera y postrera vez muy molidas, mezclándose con granos de maíz cocidos y lavados; y ansí molidas y mezcladas les echan agua en algún vaso. Si les echan poca, hacen lindo cacao, y si mucha, no hacen espuma. Y para hacerlo bien hecho se hace y se guarda lo siguiente, conviene a saber: que se cola; después de colado se levanta para que chorree, y con esto se levanta la espuma, y se echa aparte; y a las veces espésase demasiado; mézclase con agua después de molido. Y el que lo sabe hacer bien hecho, vende el cacao lindo, y tal, que solos los señores lo beben: blando, espumoso, bermejo, colorado y puro, sin mucha masa. A las veces le echan especies aromáticas, y aun miel de abejas o alguna agua ro-

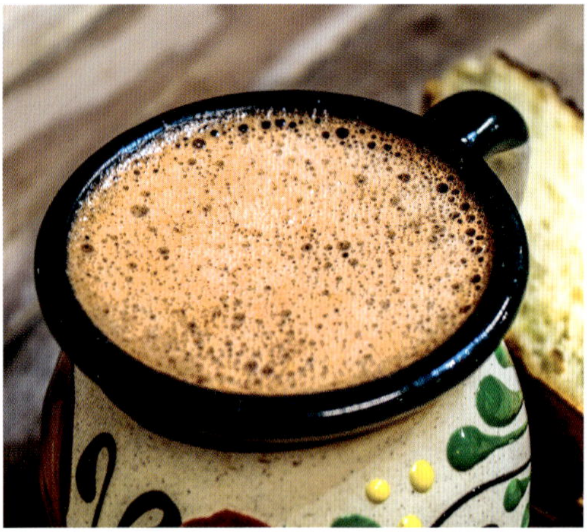

Cup of frothy chocolate.

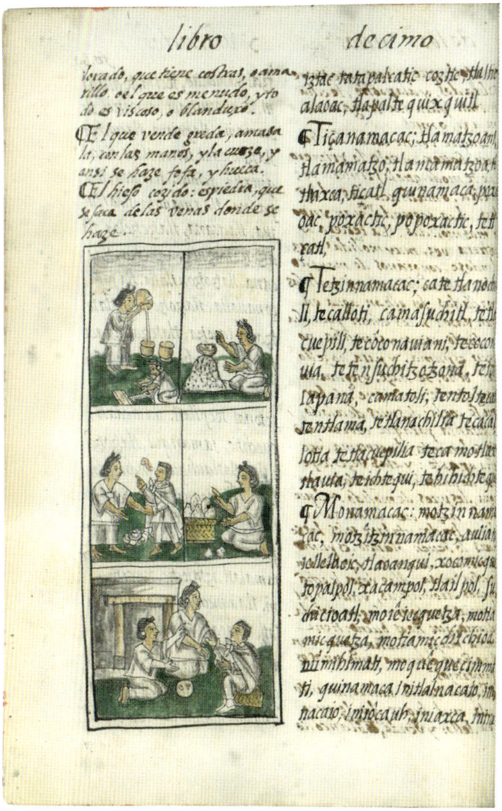

sada. Y el cacao que no es bueno tiene mucha masa y mucha agua, y ansí no hace espuma, sino unos espumarajos.

('Those who sell cocoa ready for drinking first grind it in the following way: firstly they break or crush the beans; secondly they grind them a little more; thirdly and lastly they grind them very finely, mixing them with cooked and washed maize cobs; after they are ground and mixed, they add water in a glass. If little is added, precious cocoa is obtained; if much is added, no froth is formed. To make it well, it is made and preserved in the following way: it is strained; after that it is poured from a height so that it drips, and in this way the froth is formed, which is set aside; sometimes it becomes too thick; it is mixed with water. Those who know how to do this well sell a precious cocoa, such that only gentlemen get to drink it: soft, frothy, red, colourful and pure, without too much mass. Sometimes aromatic spices are added, and even honey or rose water. Cocoa that is not of good quality has a lot of mass and a lot of water, and therefore does not form froth, but only bubbles').

Bernardino de Sahagún, Mediceo Palatino 220, fol. 71r. [R.V.]

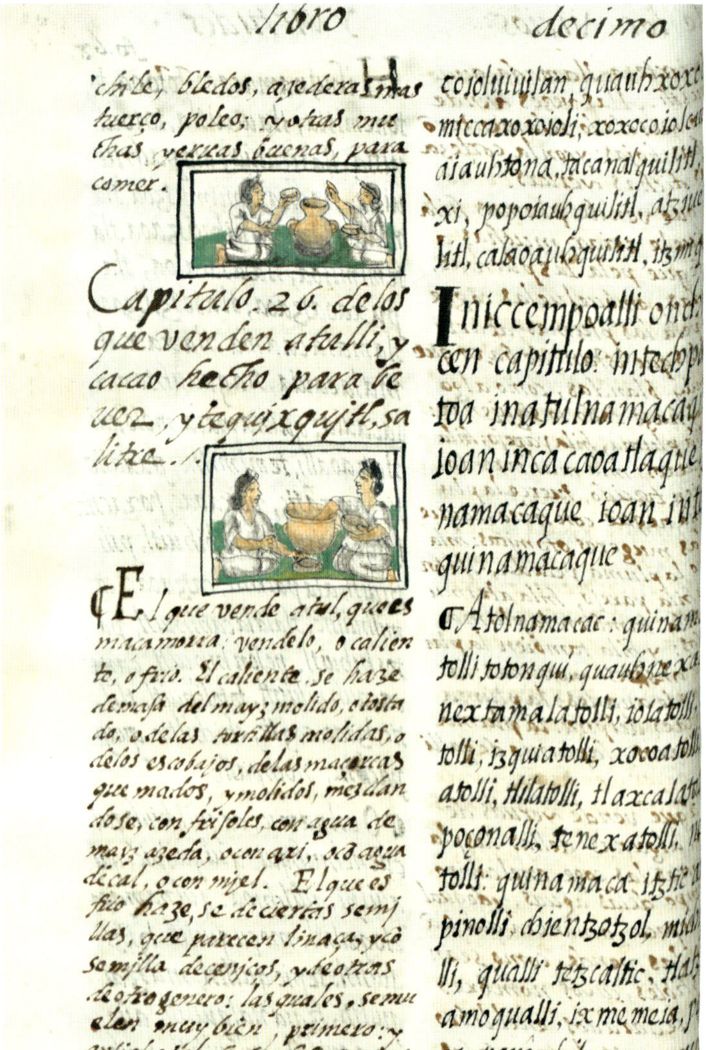

libro decimo

chile, bledos, azederas, mas tuerço, poleo; y otras mu chas yervas buenas, para comer.

Capitulo. 26. delos que venden atulli, y cacao hecho para be uer, y tequixquitl, sa litre.

El que vende atul, que es macamona: vendelo, o calien te, o frio. El caliente se haze de masa del mayz molido, o tosta do, o delas tortillas molidas, o delos escobajos, delas maçorcas quemados, y molidos, mezclan dose, con frisoles, con agua de mayz azeda, o con axi, o con agua de cal, o con miel. El que es frio haze, se de ciertas semi llas, que parecen linaça, y co semilla de cenicos, y de otras de otro genero: las quales, se mu elen muy bien, primero: y assi el atul hecho destas semi llas parece ser cernido: y quando no estan bien molidas, haze

coiolxiuitlan, quauhxox
miccacaxoxoioli, xoxocoio
aiauhtona, tzacanalquilitl,
xi, popoiauhquilitl, atzu
litl, calaoaiuhquilitl, itz

Inic cempoalli onch
cen capitulo: intechpa
toa in atulnamacaq
ioan in cacaoatlaque
namacaque ioan in t
quinamacaque

Atolnamacac: quinam
tolli totonqui, quauhnex
nextamalatolli, iolatolli
tolli, izquiatolli, xocoatolli
atolli, tlilatolli, tlaxcalat
poçonalli, tenexatolli
tolli: quinamaca itztic
pinolli, chienpotzol, mic
lli, qualli tetzcaltic, tla
amoqualli, ixmemeia,
ca, pane, chilpane, necu

Tlaquetzalnamacac

Chocolate recipe. Florence, Biblioteca Medicea Laurenziana, Mediceo Palatino 220, Mexico, 16th century (1577), fol. 70v.

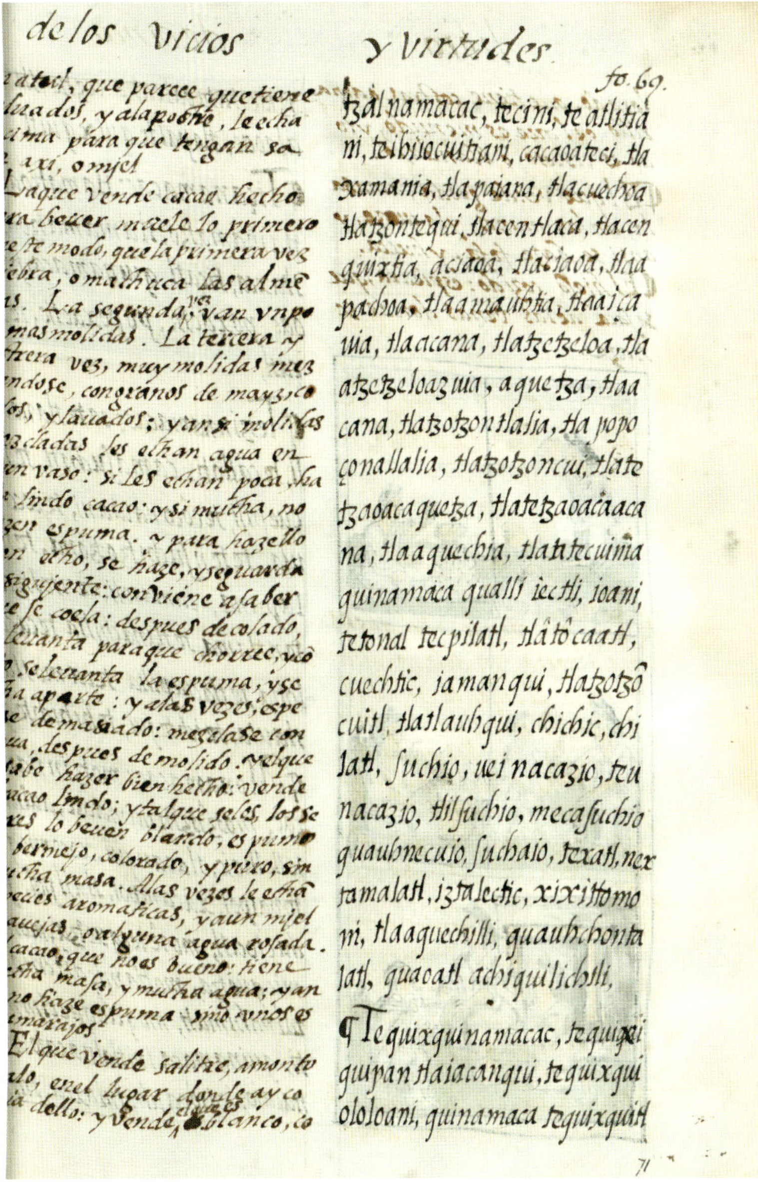

delos vicios y virtudes

fo. 69.

[illegible] atoil, que parece que tiene [illegible] lauados, y ala postre, le echa [illegible] ma para que tengan sa[illegible] axi, o miel. El que vende cacao hecho para bever, muelelo primero de este modo, que la primera vez quiebra, o machuca las almendras. La segunda: van vn poco mas molidas. La tercera y postrera vez, muy molidas, mezclandose con granos de mayz cozidos, y lauados: y ansi molidas, y amasadas les echan agua en vn vaso: si les echan poca, sale lindo cacao: y si mucha, no haze espuma. Y para hazello bien hecho, se haze, y se guarda lo siguiente: conviene a saber que se cuela: despues de colado, se leuanta para que chorree, y con esto se leuanta la espuma, y se echa aparte: y a las vezes, espesa demasiado: mezclase con agua, despues de molido: y el que sabe hazer bien hecho: vende cacao lindo; y tal que solos los señores lo beuen blando, espumoso, bermejo, colorado, y puro, sin mucha masa. A las vezes le echan especies aromaticas, y aun miel de auejas, o alguna agua rosada. El cacao, que no es bueno: tiene mucha masa, y mucha agua; y ansi no haze espuma, sino vnos espumarajos. El que vende salitre, amontonado, en el lugar donde ay copia dello: y vende, blanco, co[illegible]

tzalnamacac, tecini, teatlitiani, teihiiocuitiani, cacaoatecí, tlaxamania, tlapaiana, tlaquechoa, tlatzontequi, tlacentlaca, tlacenquixtia, aciaoa, tlatiaoa, tlaapachoa, tlaamauhtia, tlaaicauia, tlaacana, tlatzetzeloa, tlaatzetzeloazuia, aquetza, tlaacana, tlatzotzontlalia, tlapopoçonallalia, tlatzotzoncui, tlatetzaoacaquetza, tlatetzaoacaacana, tlaaquechia, tlatitecuimia quinamaca qualli iectli, ioani, tetonal tecpilatl, tlâtocaatl, cuechtic, jamanqui, tlatzotzôcuitl, tlatlauhqui, chichic, chilatl, suchio, ueinacazio, teunacazio, tlilsuchio, mecasuchio quauhnecuio, suchaio, texatl, nextamalatl, iztalectic, xixittomi, tlaaquechilli, quauhchontalatl, quacatl achiquilichtli.

¶ Tequixquinamacac, tequixquipan tlaiacanqui, tequixquiololoani, quinamaca tequixquitl

71

Chocolate recipe. Florence, Biblioteca Medicea Laurenziana, Mediceo Palatino 220, Mexico, 16th century (1577), fol. 71r.

MUJER

> Y después que había comido el señor, luego mandaba a sus pajes o servidores que diesen de comer a todos los señores y embaxadores que habían venido de algunos pueblos … Y en acabando de comer, luego se sacaban muchas maneras de cacaos, hechos muy delicadamente, como son éstos … cacao hecho de mazorca tierna de cacao … cacao hecho con miel … cacao hecho con hueinacaztli, xoxouhqui, tlilxochyo, cacao hecho con tlilxóchitl tierno … cacao hecho colorado … cacao hecho bermejo … cacao hecho naranjado … cacao hecho negro … cacao hecho blanco.
> ('After the master had eaten, he ordered his pages or servants to feed all the lords and ambassadors who had come from various countries … And when they had finished eating, many varieties of chocolate were brought forth … chocolate made from the green fruits of cacao … chocolate made with honey … chocolate made with *hueinacaztli*, *xoxouhqui*, *tlilxochyo*, chocolate made with delicate flowers … bright red chocolate … dark red chocolate … orange chocolate … black chocolate … white chocolate').
> Bernardino de Sahagún, Mediceo Palatino 219, fol. 275r.

The preparation of chocolate was the exclusive preserve of women, from all social strata: from nobility to servants. Women ground the cacao beans with a lava stone tool called *metate*; this tool was heated by a brazier underneath, while on its surface the beans were ground with the aid of a stone called *mano de metate*. The ground cocoa was mixed with water and then froth was created by pouring the drink from one vessel to another. The froth was greeted with great appreciation, demonstrating the skill and care that went into making the cocoa.

However, despite their expertise in the preparation of chocolate, Sahagún informs us that women were neither allowed to drink chocolate nor participate in the festivals where it was consumed. Sumptuary laws regulated the use and consumption of cocoa, making it an important status symbol, from which women were excluded.

The most prized versions of *cacaoatl (atlaquetzalli)* were restricted to noblemen, priests, warriors and *pochtecas*; served in richly decorated vessels, they were perfumed with flowers. Among the many species, the flowers of *xóchinacaztl* (*Cymbopetalum penduliflorum*), *tlilxóchitl* (*Vanilla planifolia*, vanilla), *mecaxóchitl* (*Piper sanctum*), *eloxóchitl* (*Magnolia dealbata*) and *yolloxóchitl (Magnolia mexicana)* were considered the choicest. [R.V.]

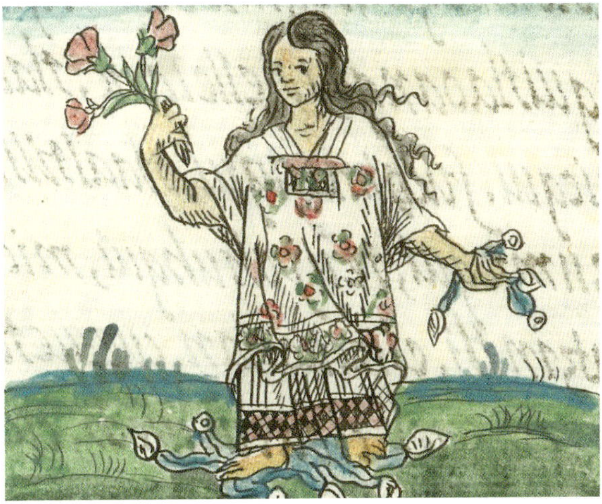

Woman with flowers. Florence, Biblioteca Medicea Laurenziana, Mediceo Palatino 220, Mexico, 16th century (1577), fol. 41v, detail.

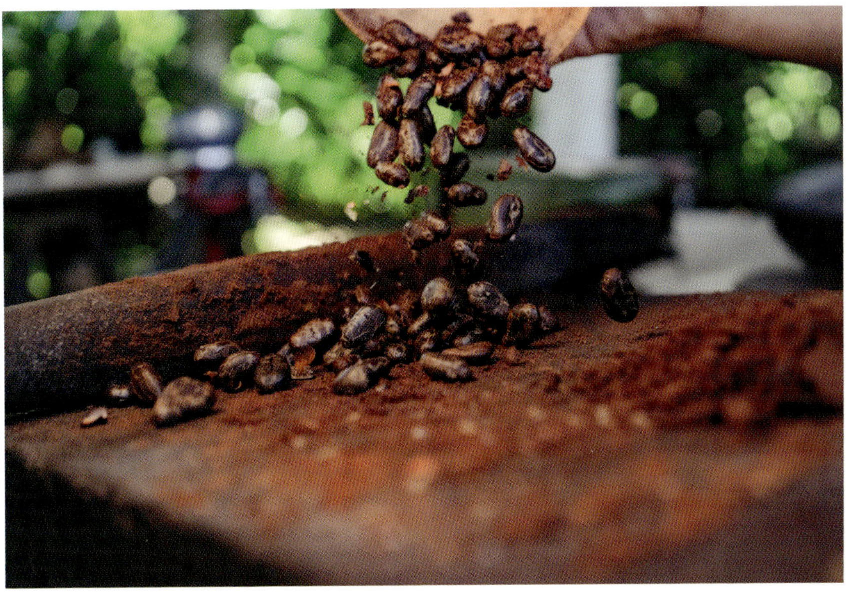

Cacao beans being ground.

METATE

The *metate* is a stone tool used for grinding grain and seeds in different cultures around the world. In particular, in Mesoamerica it was an essential element in food preparation, especially for the production of chocolate and corn tortillas. The *metate* has a long history dating back to the Upper Xenolithic period in Mexico, around 5000–3000 BC, and the Middle Archaic period in the American southwest, around 3500 BC. It consists of a large stone with a hollow or smooth bowl in the upper surface. Materials are ground on the *metate* with a *mano de metate* or *metlapil*. The grinding motion is horizontal and differs from the vertical action used in a mortar with pestle. The depth of the bowl varies according to the duration of use and the level of activity. The specific angles of the metate body allow the grains to be turned into flour effectively. There are different types of *metate*, depending on the shape of the milling surface. One classification distinguishes four shapes: flat, flat/concave, basin-shaped and trough-shaped. Basin-shaped *metates* are used with a circular stroke, while flat and trough-shaped *metates* employ a reciprocating or oscillating motion. Ceremonial *metates* made of volcanic stone are also found in Costa Rica and are among the most unusual and complex traditions of pre-Columbian artefacts. The *metate* varies between regions and historical periods. It often features zoomorphic or anthropomorphic iconography, such as birds, jaguars, monkeys or lizards. These images are thought to have a symbolic significance linked to agricultural fertility, cosmology or genealogy. [L.H.]

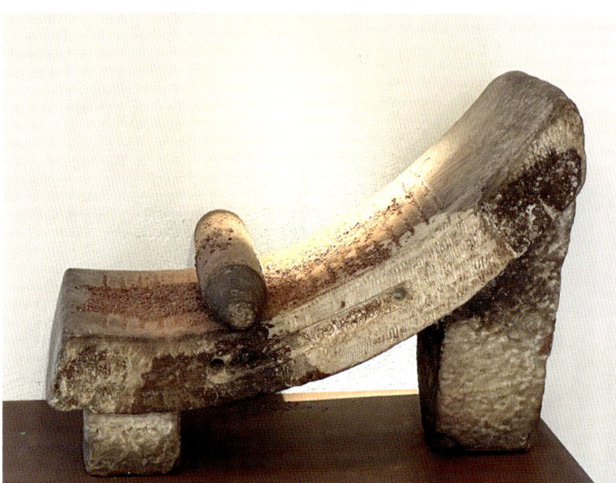

A 19th-century Mexican *metate*.

XICALLI

The vessels mentioned by Sahagún in the recipe are *jicaras*: shells of the fruits of the *xicalquahuitl* tree of various sizes. Friar Bartolomé de Las Casas (1484–1566), one of the first Dominicans in America, describes a banquet of the emperor, in which he states that the cups for drinking chocolate were not made of gold, as some *conquistadores* claimed, but rather *jicaras* painted on the inside and outside. Las Casas says that 'any gentleman could drink them as if they were made of gold and silver' (Friar Bartolomé de Las Casas, *Apologética historia de las Indias*, Madrid, 1909, p. 553). These cups, called *xicalli* in Nahuatl, were used extensively for drinking chocolate throughout Mesoamerica.

> Y dábanlo en unas xícaras con que se bebía, son de muchas maneras: una dellas se llama tecontlacuilolli, 'xícara pintada con diversas pinturas', y su atapadero muy rico, que se llama atzaccáyotl, y también su cuchara de tortuga para revolver el cacao; otra manera de xícaras se llaman ayotectli tlacuilolli, 'xícara negra pintada de negro'; y también su rodeo hecho de cuero de tigre.
> ('They served it in *jicaras* intended for consumption, of various types: one of them is called *tecontlacuilolli*, "*jicara* painted with various paintings", with a richly decorated lid known as *atzaccáyotl*, and also with a spoon for stirring the cocoa, called *atzaccáyotl*; another variant of the *jicara* is the *ayotectli tlacuilolli*, "*jicara* painted black", with a rim made of tiger skin').
> Bernardino de Sahagún, Mediceo Palatino 219, fol. 275r. [R.V.]

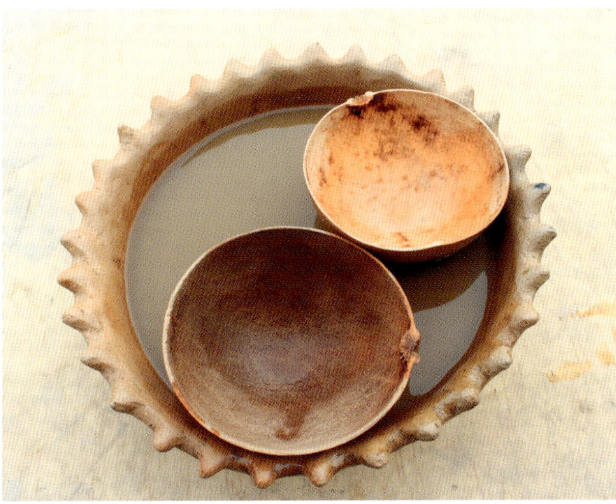

Jicaras for tasting chocolate.

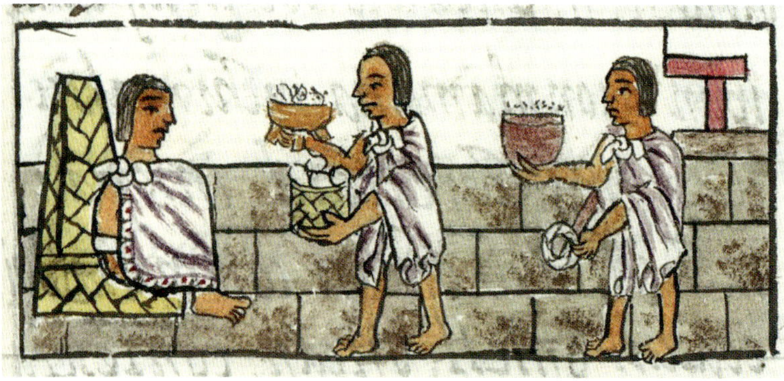

ELITE

Florence, Biblioteca Medicea Laurenziana, Mediceo Palatino 219

Mexico, 16th century (1577)
fol. 337r, detail

Among the Mexica and other Mesoamerican peoples, drinking chocolate was a privilege reserved for the elite: the royal family, rulers and the noblemen. The only commoners who had the chance to savour such luxury were warriors and *pochtecas*. The latter were regarded as soldiers because they were armed on their journeys through dangerous lands to distant markets. Priests, high-profile figures in Mesoamerican society, did not consume chocolate as they were supposed to lead lives of extreme austerity and penitence.

16th-century chronicles, recorded not only by Bernardino de Sahagún, but also by Diego Durán, Gonzalo Fernandez de Oviedo y Valdés, describe scenes of ritual banquets both in the Central Plateau, where the Mexica lived, and in Yucatán, the land of the Maya, during which copious amounts of chocolate were consumed. At banquets and meals attended by the elite, chocolate was not drunk with the food, but was served at the end, along with tobacco.

Chocolate beverages enjoyed the same status as expensive French champagne in today's Western culture. Ethnohistorical accounts indicate the extensive use of chocolate in betrothal and wedding ceremonies, and even in those involving children, especially among the wealthier classes. [R.V.]

SANGRE

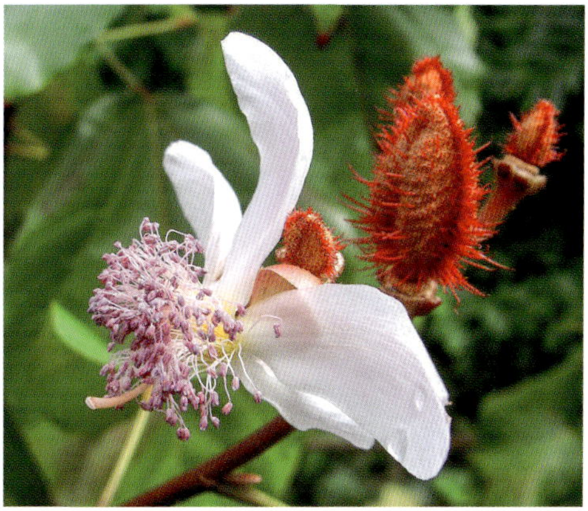

Biza orellana.

The cacao tree had deep symbolic significance for Mesoamerican civilisations. It is the Tree of the South, the direction of the Land of the Dead, associated with the colour of blood. A variant of *cacaoatl* was coloured with the flower *Biza orellana* (Achiote), which gave chocolate an intense red shade like blood. Various metaphors were used to define cacao, including 'yolotl, eztli', which can be translated as 'heart, blood'.

> In one spectacular ritual that took place every year in Tenochtitlán, cacao was directly associated with heart extraction. Once a year, a slave perfect in body was chosen to impersonate the great god Quetzalcóatl. For 40 days, he went around dressed in the raiment and jewels of the god, during which time he was treated as the god himself … on the eve of his sacrifice, he was warned by the temple elders that he was to die, after which he had to perform a dance, which had to be carried out as though he were completely happy about his fate. And if he were not? … they [the priests] went immediately to procure sacrificial knives, washed off the human blood adhering to them (the result of past sacrifices), and with that filthy water prepared a gourd of chocolate, giving it to him to drink. It is said that the draught had this effect upon him: he became almost unconscious and forgot what he had been told. Then he returned to his usual cheerfulness and dance … It is believed that he offered himself for death with great joy and gladness, bewitched by the beverage. This drink was called *itzpacalatl*, which means 'water from the washing of obsidian blades'.
> Sophie D. Coe, Michael D. Coe, *The True History of Chocolate*, London, 1996, p. 102.

This symbolic blood-related significance of chocolate may have been transferred to the military sphere, explaining perhaps its inclusion in military rations for the army. Being a symbol of blood, it was the most fitting drink for Aztec warriors.

> Y luego diéron los armas para conque peleasen. Y diéronlos mucho cacao. Y luego los pusieron en el lugar donde habían de pelear.
> ('Then they gave them weapons to fight with, and they gave them much cocoa, and then they put them in the place in which they were to do battle').
> Bernardino de Sahagún, Mediceo Palatino 220, fol. 469r. [R.V.]

CHOCOLATE

Cortés always used the word 'cocoa' to designate chocolate, which the Spanish acquired from the Maya of Yucatán and Tabasco. At some point in the second half of the 16th century, the Spanish introduced a new word: 'chocolatl'. The word may have come from the root 'xoco', meaning bitter, and 'atl', meaning water, thus forming 'xocoatl'. Another possible source of the word 'chocolate' could be the Maya-Quiché word 'chokola'k', meaning 'drinking chocolate together'. However, philologist José Ignacio Dávila Garibi puts forward another hypothesis: the Spaniards adopted the Maya word 'chocol', meaning hot, and the Nahuatl word 'atl', meaning water, thus forming the word 'chocolatl'. This choice of name may have been motivated by the need to avoid the original word 'cacaoatl' because of the presence of the two syllables 'caca', since in Spanish and Latin they correspond to the vulgar word for excrement. It is conceivable that the more cultured friars proposed the term 'chocolatl', which then evolved into the current form of 'chocolate'. [R.V.]

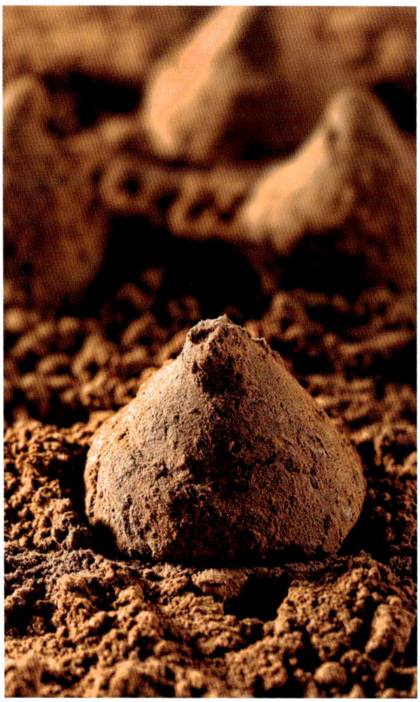

Cocoa powder.

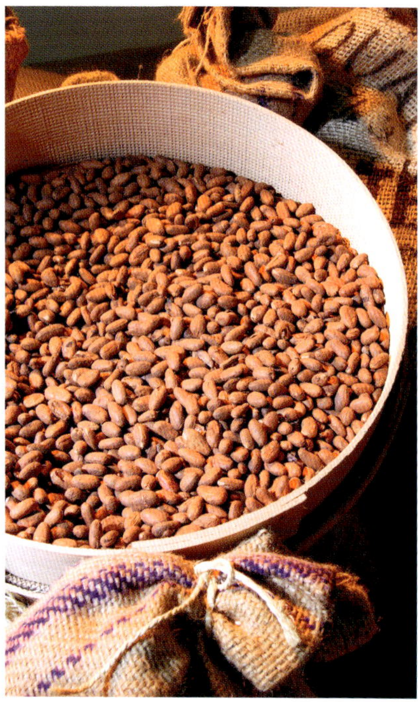

Cacao beans.

VIRTUD

As the renowned German explorer and naturalist Alexander von Humboldt wrote, 'Nowhere else has nature compressed such a store of the most valuable nutrient into such a small space as it has with the cacao bean'.

Cacao was a medicinal remedy used in pre-Hispanic times without any restrictions in terms of gender or social class. In the chapter dedicated to 'medicinal herbs' in Book X of Sahagún's work, cocoa preparations are mentioned as being used to treat 'internal heat and faintness of heart', as well as to hasten childbirth, clarify urine, relieve coughs, open pores and improve digestion.

Today, cacao beans are still recognised for their special pharmacological properties, which include neuroactive, antioxidant and stimulant elements. More than 400 chemical elements have been identified in their composition, including three of the most-studied alkaloids: theobromine, caffeine and theophylline. Cocoa stimulates the production of serotonin and endorphins, the good mood neurotransmitters, producing an antidepressant effect.

Containing more phenolic antioxidants (flavonoids above all) than most foodstuffs, it has one of the highest antioxidant strengths of all: it can help defend us against cellular ageing, tumours, diseases such as Alzheimer's and protect nerves from injury and inflammation. It directly impacts on insulin resistance and, consequently, reduces the risk of diabetes. According to a study conducted by the University of Pennsylvania, it is able to reduce blood cholesterol levels by up to 10 per cent and research by the Queen Mary's School of Medicine and Dentistry, London, has shown that cocoa has a preventive function against heart attacks, thanks to epicatechin, a flavonoid that thins the blood. It lowers blood pressure levels, improves vein and artery health and decreases the amount of LDL (the 'bad' cholesterol), while increasing HDL (the 'good' one).

A scientific investigation by Wheeling University, USA, reports that cocoa apparently improves visual and verbal memory, reaction time, attention, problem-solving ability and impulse control.

A recent article in *Jama* (*Journal of the American Medical Association*) has refuted the link between chocolate consumption and acne. Cocoa is able to protect the skin from the oxidative damage caused by UV radiation in topical preparations.

Organoleptically, it contains various minerals and trace elements, including magnesium, calcium, potassium, iron and selenium, which are important for maintaining healthy muscles, bones and thyroid. 'Criollo' cacao is unquestionably the best in terms of beneficial qualities. To obtain the best benefits of the properties of cocoa, dark chocolate is the ideal choice, as long as it is at least 70% pure and free of excessive added sugars and fats.

It is also interesting to note that in the holistic context, cocoa takes centre stage in ceremonies involving mind and spirit, aimed at creating unique multi-sensory experiences. At the opposite end of the spectrum, it is widely used in industry in goods and services geared towards improving personal appearance and aesthetic care. [L.H.–R.V.]

5. THE VANQUISHED

Book XII, Conquest

'de la conquista de la Nueva España, que es la ciudad de México'

'En el libro XII de la *Historia general*, el vencido pasa a ser protagonista del relato en su propia lengua; por ello este libro es considerado paradigma de una nueva categoría histórica, la de "la visión de los vencidos"'

('In Book XII of the *Historia general*, the vanquished become the main characters in this story told in their own language; for this reason, this book is considered a prototype of a new historical category, that of the "vision of the vanquished"').

Ascensión Hernández de León-Portilla, 'La historia general de Sahagún a la luz de las enciclopedias de la tradición greco-romana', in *Bernardino de Sahagún: quinientos años de presencia*, Mexico, 2002, pp. 41–59: 58

The concept of 'vision of the vanquished' comes from the title of a collection of short stories about the Conquest of Mexico published by Miguel León-Portilla in 1959.

Florence, Biblioteca Medicea Laurenziana, Mediceo Palatino 220

Mexico, 16th century (1577)
fol. 465r

The 12th book, known as the 'Libro de la Conquista', is notable for being the first chronicle of this enterprise written from the perspective of the vanquished. Sahagún states that he compiled it with the intention of incorporating Nahuatl vocabulary related to the militia; however, the methodological and linguistic value of this work diminishes as one enters into the narration of events. The Mexican poet and essayist Alfonso Reyes (1914–98) considers Book XII a gem, as it offers 'the comparison of the same facts from divergent perspectives, that of the conqueror and that of the conquered' (*Letras de la Nueva España*, México, 1948, p. 54).

Most of Sahagún's work was completed during the reign of Philip II (1527–98). As soon as it was completed, the work, the only source of cultural enlightenment in the darkness of the conquest, was quickly suppressed by the direct intervention of the King of Spain himself: on 22 April 1577, Philip II issued a 'Real cédula de prohibición', forbidding anyone to access Sahagún's work or to contribute to its dissemination, ordering the immediate confiscation of all the volumes. [R.V.]

el monte que se llama tepetzinco, donde estan los baños.

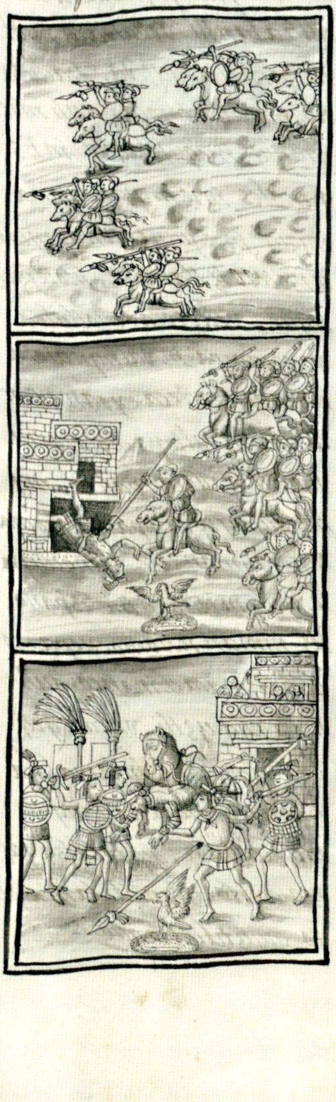

cavan. Auh in Españoles, qi
njman amo mamana: auh
in oquj ttazque tlequjqujztli
cenca tlaiovac, poctli moteac
Auh in qujmotocti timanca
in tetlaquetzalli, netlaloloc
yoan in jxqujchtin tlapan
co onoca valmotepeuhque
vecanetlaloloc, necquj valle
cavia in tlequjqujztli con
tecaco temalacatitlan: auh
in jicpac vitzilobuchtli, ocne
tlapiaia, qujvitequja in tepo
naztli, iuhqujn ontetemjquj
vitequj teponaztli: auh njmã
iconteleoque vmentin Españo
les, vmpa qujmonvivitequito:
auh in oqujmonvivitequito in
cavalmaiauhque, qujn valehi
canauhque: auh in quaquachi
tin in jxqujchtin intiacavan
in acaltica tlaiecoaia, muchin
tin valqujzque, vallalvaca
qujzque: auh in tetlanehuja
in telpopotzitzinti, qunieho
tin in qujvalhujvicaque in

CEDULA

Jacopo da Trezzo, *Bust of Philip II of Spain*

c. 1559–62
Onyx, engraved gold and polychrome *champlevé* enamel, 42 × 33 mm
Florence, Gallerie degli Uffizi, Palazzo Pitti, Tesoro dei Granduchi, inv. Gemme, no. 127

> Su Majestad el Rey.
> To Don Martín Enríquez, our Viceroy, Governor and Captain General of New Spain, as well as President of our Reale Audiencia (appellate court) in those lands. We have received some letters from those provinces informing us that Friar Bernardino de Sahagún, of the Order of St Francis, has produced a universal history of the most remarkable things in New Spain. This work is a detailed account of all the rites, ceremonies and idolatries that the natives practised in their unbelief. It is divided into twelve books and written in the Mexican language. Although we understand that the intention of the said Friar Bernadine was noble and that his work is intended to be beneficial, we have reached the conclusion that it is not appropriate for this book to be printed or distributed in any way in those regions. Therefore, we command you that as soon as you receive this decree, with great care and diligence, you ensure that you obtain these books. Without overlooking any copies, you must send them at the first opportunity to our Council of the Indies. Furthermore, we order you not in any way to permit anyone to write about the superstitions and ways of life practised by these natives, in any language, since this is necessary for the service of God our Lord.
> Madrid, the twenty-second day of April, one thousand five hundred and seventy-seven.
> Yo el Rey.

Philip of Spain, after the abdication of his father, Emperor Charles V, inherited the Spanish crown in 1556. In the first twenty years of his reign, Philip proceeded to consolidate the Spanish territories, aware that the challenges of Hispanic colonization required a profound understanding of the Amerindian world so that it could be addressed and dealt with. However, from 1575 onwards, after the death of the President of the Council of the Indies, Juan de Ovando (1530–75), Philip II's policy took a radical turn and he decided to brutally extirpate everything Amerindian: languages, religions, cultures and identities. The obsession with the purity of Catholicism imposed on the new indigenous subjects could under no circumstances tolerate their socio-cultural past. A reading of the king's orders, in the Cédula Real, reveals the origin of letters from Mexico denouncing Sahagún's work, as being dangerous because it preserved the memory of idolatrous beliefs and rituals.

Sahagún obeyed and delivered the twelve books into the hands of his patron, Father Rodrigo de Sequera.

Friar Sahagún, being extremely old, did not clearly understand the scope of these orders and even thought it appropriate to write personally to Philip II, on 28 March 1578,

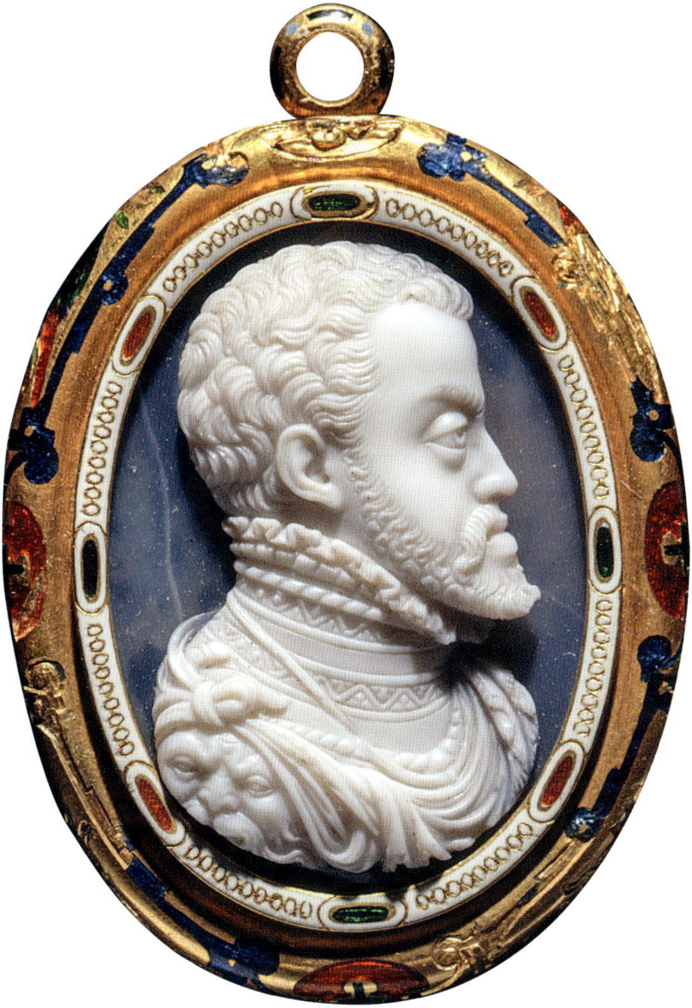

to confide in him that he had kept most of his notes and could provide him with further copies if he wished:

> Viceroy Don Martín Enríquez received a Decree from Your Majesty ordering him to send Your Majesty some works I wrote in Mexican and Spanish. The Viceroy informed me of this, as did the Archbishop of this city. I completed and put in order all these works last year and gave them to Brother Rodrigo de Sequera, Commissary General of our Order of St Francis, so that, if he were to go (to Spain), he might bring them to Your Majesty, and if not, that

> he might send them, for when the decree arrived, he already had them in his possession. I have heard that the Viceroy and the Commissary were to send to Your Majesty these works divided into twelve books, in four volumes, on the ship that departed recently. If this is not the case, I humbly beg Your Majesty to order that I be informed, so that a new shipment might be prepared and this opportunity not be lost, to avoid the memorable things of this New World from falling into oblivion. He who delivers this letter, the Custodian of this Province, is on his way to the General Chapter; if Your Royal Majesty so desires, you may obtain information about me and my works from him.
> Bernardino de Sahagún, Letter to Philip II, Mexico, 26 March 1578.

The friar had naively believed that the King and the Council of the Indies were interested in his writings to learn 'about the things of New Spain'. In the years that followed, Bernardino never heard anything more about the fate of his books, which in 1585 he described as highly 'historiated' (profusely illustrated):

> The books, of which there were twelve, were sent on behalf of our King Don Felipe, and I sent them … Afterwards, Father Rodrigo de Sequera took them away them after fulfilling his duty as Commissary in this land, and he has never written to me about what happened to those books he took with him in Mexican and Spanish, very rich in history; nor do I know now in whose hands they are.

This must doubtless have caused him great concern and distress, but not discouragement. Despite being over eighty years old, he continued to write to enrich his remaining texts. Bernardino de Sahagún died in 1590 in Tlatelolco, never knowing what fate had befallen his books.

In Sahagún's work, the voice of the Vanquished speaks – those who were reluctantly part of the greatest socio-economic revolution in history: that revolutionary process of exchange *entre dos mundos*, between two worlds, of men, plants, animals, artefacts and diseases. In this respect, it might also be considered a small payback on their part that cocoa, in the form of chocolate, then conquered the Old World. [R.V.]

Attributed to Johannes Vingboons and workshop
The Port of Veracruz with the fortress of San Juan de Ulúa in Nueva España in the 'Mar del Norte'

Amsterdam, 1667
Line-and-wash drawing on canvas-backed paper, 60 × 47.2 cm
Florence, Biblioteca Medicea Laurenziana, Carte di Castello, 41

The Olmec civilisation, which evolved in the lands south of Veracruz towards the end of the second millennium BC and today is called the 'mother culture' of Mesoamerica, extended northwest of Tabasco and existed until 400 BC. The cult of the Plumed Serpent, adopted by many later cultures, probably originated in Veracruz, in the area of Coatzacoalcos, whose name means 'place where the serpent hides', precisely where, the legend goes, Quetzalcóatl fled. In pre-Columbian times Veracruz was called Totonacapan, the land of the Totonacs and when the Spanish landed there, on Good Friday 1519, the day of the True

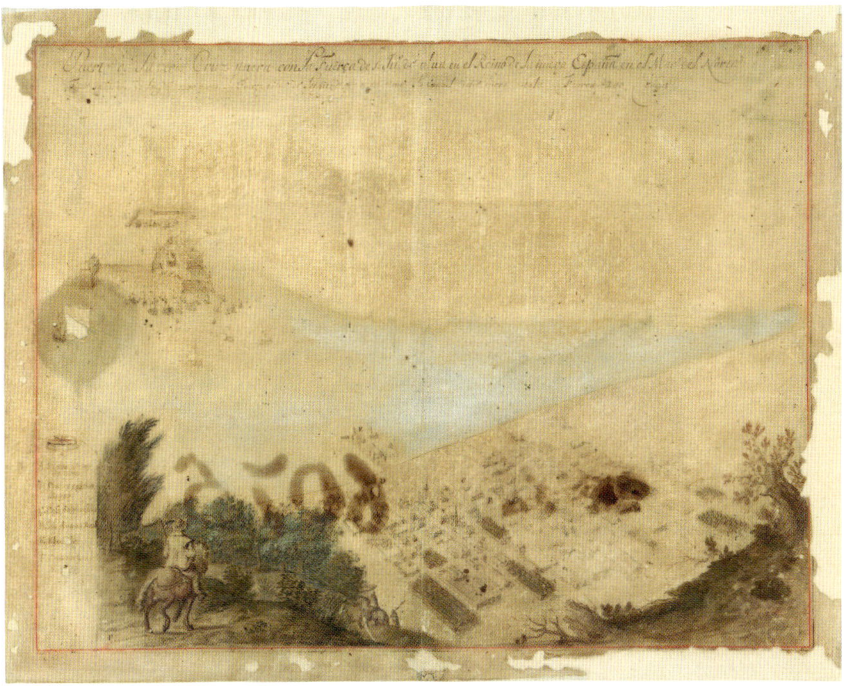

Cross, this place was given the name of Villa Rica de la Vera Cruz. It was from here that Hernán Cortés set off for the city of Tenochtitlán to conquer Mexico.

From the 16th century onwards, Veracruz established itself as the main port connecting the New and Old Worlds, contributing to the establishment of the first global trade exchange network between Europe, China, Africa and America. The spread of previously isolated biotic organisms was triggered by these exchanges, a phenomenon subsequently referred to as the 'Columbian Exchange'. The main impact of this event, the globalisation of food, would permanently revolutionise food worldwide. In one part of the world there was the arrival of wheat, rice, pigs, cows, horses, lambs, citrus fruits, grapes, onions, sugar cane, bananas, olives and coffee; in the other part, tomatoes, maize, potatoes, beans, vanilla, avocados, peanuts, chilli peppers, tobacco, pumpkins, courgettes, turkey, pineapples and cacao.

The cacao beans were passed along the lines of communication between the American convents and their respective mother houses in Spain. The first cacao beans, coming from Veracruz, arrived in Seville between 1521 and 1532, in the monastery of the third order of Franciscans. According to other sources, it was Friar Gerónimo de Aguilar (1489–1531), the shipwrecked Spanish interpreter of Cortés, together with Malinche (1502–1527/9), who introduced chocolate to Europe by sending sacks of cacao beans together with a recipe for making chocolate to the Monasterio de Piedra (Aragon, 1534). Others attribute the introduction of cacao to a delegation of Dominican friars who, upon their return from Mexico in 1544, led a group of Kekchi Maya dignitaries on a visit to Philip of Spain. The guests offered the prince many gifts, including a mysterious dark drink.

The first commercial shipment of cacao beans in Europe dates back to 1585, to Seville from the port of Veracruz. This was the beginning of the trans-oceanic cocoa trade, for which the Spanish would long hold a monopoly, facilitated by direct access to the product. The indigenous people were recruited as slave labour through the *encomienda*, a system administered locally by landowning Spanish lords who were mostly ex *conquistadors*. The colonies were expected to supply gold, silver, and agricultural products to the homeland. It was in this context that cacao was grown, consumed and traded. [R.V.]

Indigenous handicrafts, *Carved wooden whisks*

Oaxaca, 20th century
donated by Zoila Vargas Arriaga

In Mesoamerica, the consumption of chocolate held a sacred importance and was a privilege reserved exclusively for figures of high social standing. When the Spanish began to consume it in the colonies and imported the drink back home, this practice persisted, becoming a symbol of prestige reserved for the social and religious elite. However, the chocolate drink that the natives prepared was too bitter for their taste. Although sweeteners such as honey, agave syrup and other substances existed in the diet of the Maya and Mexica, they were not enough to satisfy the Spanish desire for sweet flavours. At this point, a process of hybridisation began through the addition of sugar and cinnamon, the origin of which remains uncertain. On the one hand, it is known that cinnamon arrived in Mexico from the distant island of Ceylon or Sri Lanka at the end of the 16th century, and it is clear that the introduction of sugarcane crops in America by the Spanish could have made the initial mixing of both products easier there. On the other hand, there are documents which suggest that this process only occurred with the arrival of cacao in Europe. In any case, this phenomenon is closely linked to the work of members of religious orders, notably a convent in Oaxaca, Mexico, and the Monasterio de Piedra in Zaragoza, Spain.

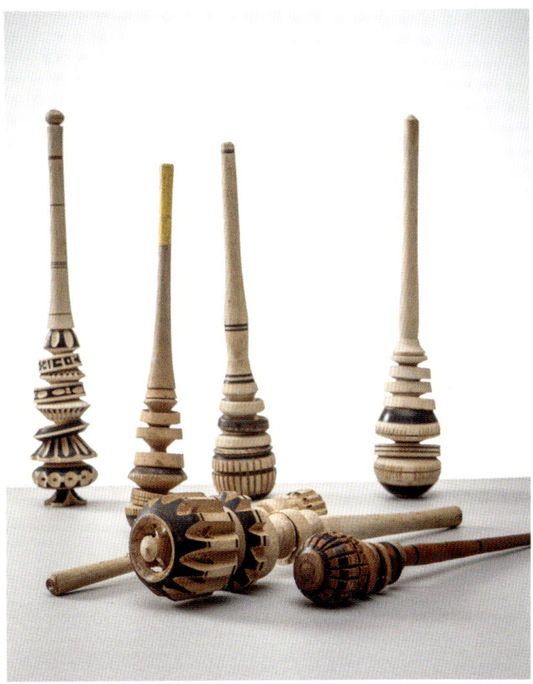

Spain, however, did not change the way in which raw cacao was prepared and turned into chocolate. The indigenous people continued to do all the work of harvesting the cacao pods and fermenting, drying, cleaning and roasting the beans on the *metate*. The technique of 'cold-processing' cacao, which originated in Mesoamerican cultures, later spread to Europe thanks to the Spanish colonists. The resulting paste was formed into small bricks or tablets – known as 'bolli' or 'bogli' – to be melted in a container with hot water and then worked into a froth.

The froth in the chocolate, with its strong spiritual significance, was widely appreciated by the Mexica and was also enjoyed by the Spanish. The common practice to produce the froth was the pouring technique, which consisted of continuously pouring some of the liquid from one vessel into another until the froth formed. However, it is likely that tools made from branches and roots were also used. The *criollos* (Creoles: the people born in Mexico from Spanish descent) modified the indigenous method of producing froth with the introduction of the *molinillo*.

The *molinillo* is a wooden whisk with a long handle and a ball-shaped tip carved into grooves and blade shapes; this tip is dipped into the hot drink and rotated between the hands to whisk the liquid and create froth. Miguel León-Portilla (1926–2019) has shown that the term 'molinillo' probably derives from the Nahuatl 'molinia', meaning 'to shake, move', and from 'moliniani', meaning 'something that moves or shakes'.

Even today, many Mexican and Spanish people use traditional tools such as *manos*, *metates* and *molinillos* to prepare cocoa-based drinks. [R.V.]

Du Paquier Porcelain Manufactory, *Mancerina*

Austria, 1740
Porcelain, decorations in black and gold, 3.5 × 12.1 × 17 cm

The full acceptance of the Mesoamerican drink involved its integration into the European medical framework, which was based on the humoral theory of disease and nutrition, originally formulated by the classical Greeks. This theory persisted until the advent of modern medicine in the first half of the 19th century. Philip II in 1570 sent his personal physician, Francisco Hernández de Toledo (1515–87), to Mexico. In the texts collected in the volume *Historia Natural de la Nueva España*, he classified the plants of Mexico as 'cold' or 'hot' and 'wet' or 'dry'. According to Hernández, the cacao bean is 'temperate in nature' but leans towards 'cold or wet', making it very nutritious as a whole. Due to its 'cool' nature, cocoa-based drinks are recommended in hot weather and to treat fevers. By adding spices (of a warm nature), it 'perfumes the breath, fights against poisoning, and relieves intestinal pain'. In another treatise on food *Primera parte de los problemas y secretos maravillosos de las Indias*, by Juan de Cárdenas (1563–1609), published in 1591, it is stated that chocolate 'makes people happy and strong'.

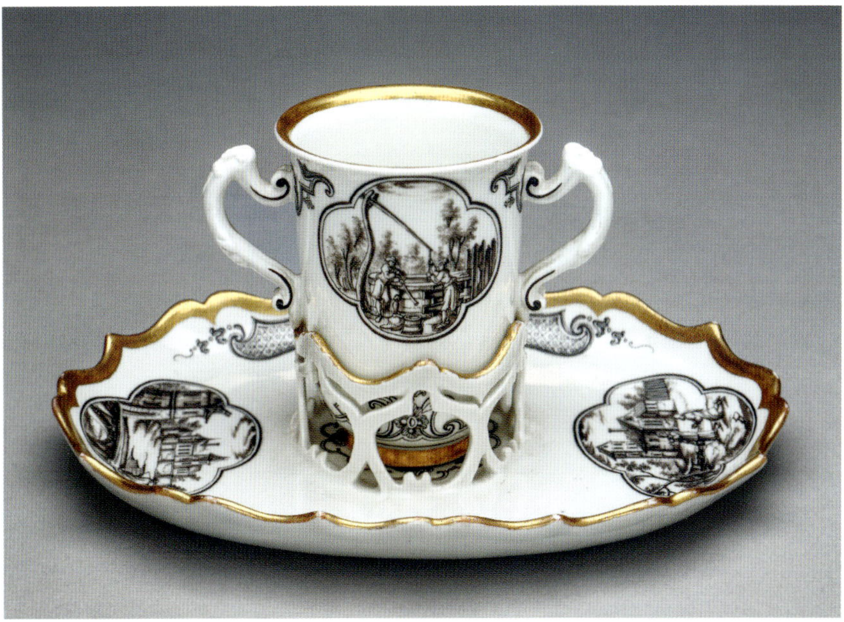

In Spain, chocolate loses its spiritual meaning but it acquires a medicinal and remedial importance for the humoral system.

Spanish royalty and nobility probably enjoyed chocolate in the traditional Mesoamerican form, served in natural or ceramic *jicaras*. However, it could not have been easy to consume with the delicacy required by courtly etiquette. It was felt that it was essential to create an *ad hoc* service in order to taste the chocolate with the proper degree of grace and elegance. The solution came from Peru: the *mancerina*.

Like the elaborate crockery of the ancient Maya and Mexica, these dishes, called *mancerinas*, were not only tableware, but were also symbols of opulence.

Around 1640, the Viceroy of Peru, Pedro Álvarez de Toledo Marquis of La Mancera (1585–1654), a refined and elegant man, would offer all kinds of choice dishes at his receptions, including hot chocolate, of Mexican origin, served in cups. Legend has it that the viceroy suffered from a tic in his hand, meaning that he constantly spilled the chocolate when drinking it. Given the high cost of this delicious food, this was literally throwing money away. Another story tells of the displeasure he felt at seeing the ladies of the court in moments of distraction staining their expensive gowns with chocolate as they drank it. So, the marquis devised a system to prevent such accidents: based on an egg cup of medieval origin: the new utensil featured a kind of plate with a circular clamp in the centre, to hold the cup firmly in place. This original solution was named after him, and initially the *mancerinas* were made of precious silver. With the spread of chocolate consumption throughout Europe at the beginning of the 18th century, the prestigious ceramics factory of the Count of Aranda, in the Alcora near Valencia, was among the first to produce them in elegant porcelain.

After the Spanish first came into contact with chocolate in the early 16th century, having overcome their initial mistrust, they jealously guarded its existence as if it were a secret for almost a century. Only the king, princes, nobles, and members of the clergy could afford the privilege of enjoying that delicious drink. However, such a delightful secret couldn't be hidden forever. During the Renaissance and the Baroque era, it established itself as a drink and a medicine, spreading to the aristocratic and elite circles of Europe because of its therapeutic qualities and the fact that, as a drink, it did not break the ecclesiastical fast, as stated in 1664 by Cardinal Francesco Maria Brancaccio (1592–1675): 'Liquidum non frangit jejunum'. [R.V.]

... and so it is reduced to a paste, which when dissolved in water becomes a drink, and is commonly drunk by all the natives of the country; and the Spanish, and every other nation that goes there, once they get used to it, become so addicted to it that they have a hard time not drinking it every morning, or late in the day after dinner when the weather is hot, especially when sailing. For this reason, it is carried in tins, mixed with spices, or made into small blocks, which dissolve immediately when put into water; and they put the water in bowls made from large fruits, which grow on some trees in those countries and are like small marrows, except that they are round and become like wood when dry. In these they drink the chocolate, and stirring it with a stick and twisting it with the palms of their hands, they make a red froth, which they immediately put into their mouths and swallow in one gulp with admirable relish and satisfaction for its nature, which gives strength, nourishment and vigour in such a way that those who are used to drinking it cannot keep themselves strong without it, although they would take things of greater substance; and they seem to lose strength when they do not have the drink. And I experienced this while I was in Mexico, I drank the chocolate, and I liked and enjoyed it very much, and I almost thought I could not go a day without taking some.
Francesco d'Antonio Carletti, *Ragionamenti ... sopra le cose da lui vedute ne' suoi viaggi si dell'Indie Occidentali, e Orientali come d'altri paesi*, Florence, in the printing house of Giuseppe Manni, per il Carlieri all'Insegna di S. Luigi, 1701, pp. 92–3.

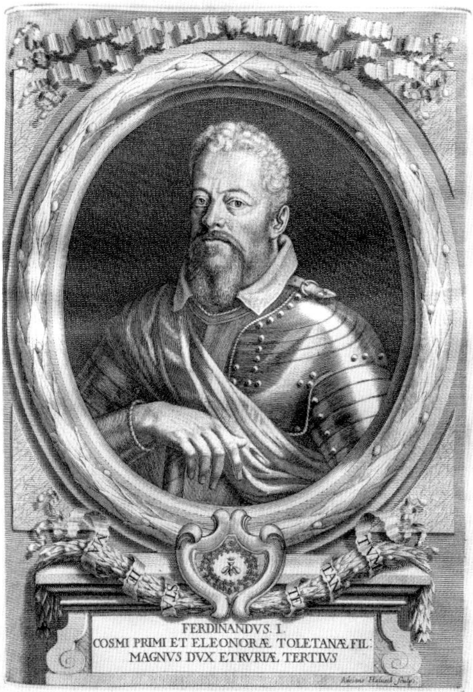

Adriaen Haelwegh, *Portrait of Ferdinand I de' Medici*, Holland, second half of the 17th century, engraving, 354 × 248 mm.

After Spain, Italy is the country where chocolate became most widespread. Some say it was brought to Italy by Filiberto di Savoia (1528–80) upon his return from a visit to Spain; others claim it was Catherine of Habsburg (1567–97) who brought it with her from Spain after her marriage to Charles Emmanuel of Savoy; all of whom are figures linked to the Medici dynasty. The most widely accepted theory is the one established by Francesco Redi (1626–97) that the first was the Florentine Francesco Carletti (1573–1636), master of the Grand Ducal House of Medici, who returned to Italy in 1606 after a long journey around the world.

Francesco d'Antonio Carletti, a Florentine merchant, set out in 1591 to travel the world in search of new markets and new products (probably as a slave trader). Near Guatemala, in 1600, he saw cacao plantations and in his travel journal he gave a detailed description of the cultivation and production by the natives of a particular drink called 'chocolate', including the use of the *molinillo*. Notable among the many stops documented in the section on the West Indies is the impactful experience in Mexico City. During his stay in this city, Carletti tells us that he had made daily use of the 'chocolate'.

Upon his return to Florence in July 1606, he shared the experiences gathered during his travels with the Grand Duke of Tuscany, Ferdinando I de' Medici (1549–1609). Subsequently, he was commissioned to oversee an ambitious project, which unfortunately was not completed, to establish an autonomous trade between Livorno and the East. During this period, Carletti completed the definitive drafting of the *Ragionamenti*, whose manuscript he delivered to the Medici. However, this work was not published until 1701 posthumously. Francesco d'Antonio Carletti received recognition from Redi for the achievement of introducing chocolate to Florence. Francesco Redi and Lorenzo Magalotti (1637–1712), eminent scholars and men of culture with a shared a passion for the delicious drink, were active members of both the scientific Accademia del Cimento, located in the Palazzo Pitti, and the Accademia della Crusca. They collaborated in the drafting of the third edition of *Vocabolario* della Crusca (1691), where the entry 'cioccolate' is written with the notation 'It is also more vulgarly called cioccolata'. Redi himself in his annotations to *Bacco in Toscana* (1685) records the variants 'cioccolate', 'cioccolatte', 'cioccolata' and 'cioccolato'.

At the Medici court, chocolate was given innovative combinations and new aromatic flavours. [R.V.]

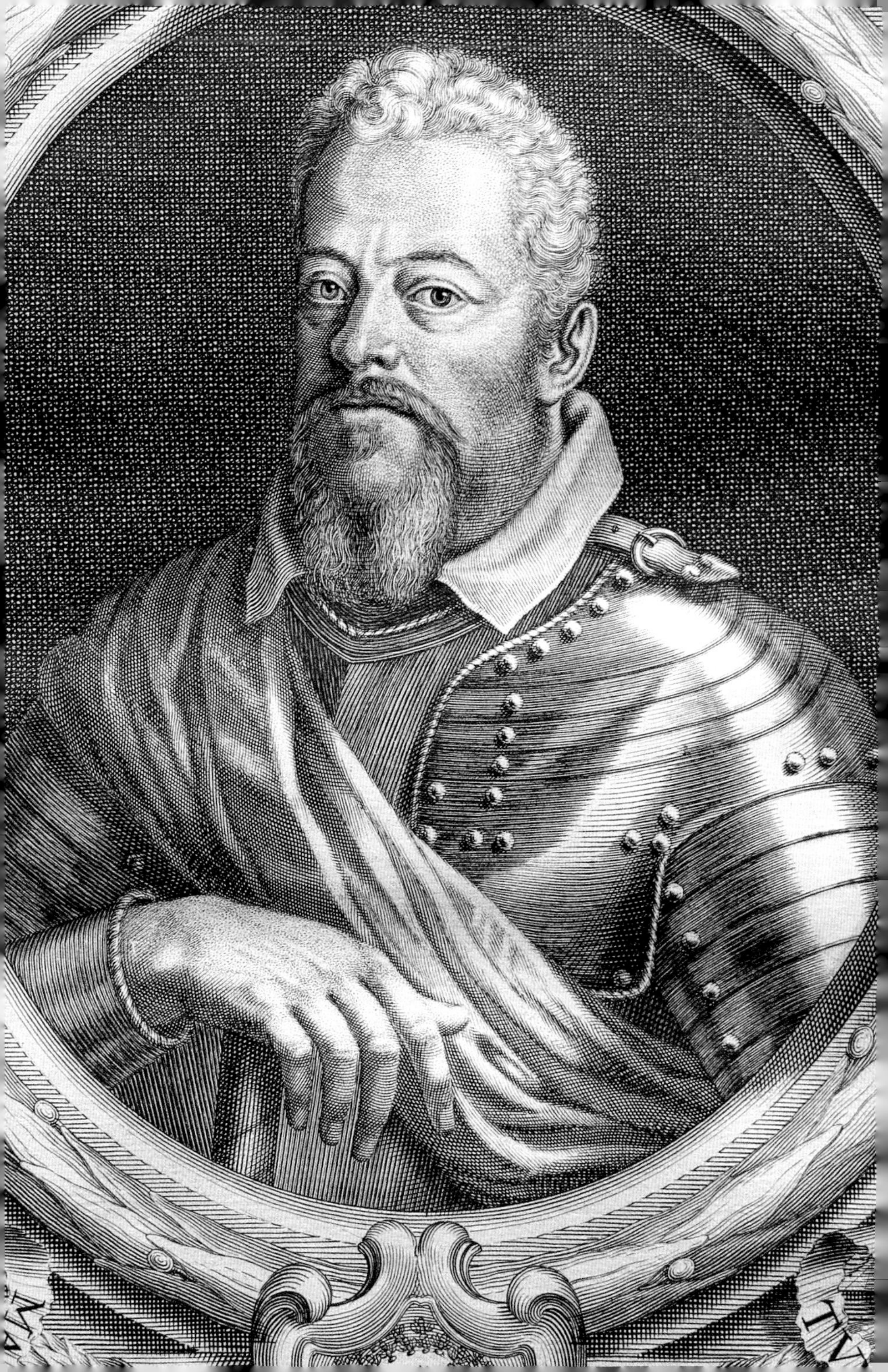

Antique French silver chocolatier with embossed decoration, with feet, boxwood handle and hinged lid

France, late 19th century
Crafted silver and boxwood, 14 × 15 cm
Trento, Collezione Roberto Caraceni

After its arrival in Italy, while the Enlightenment was rising from the ashes of the Renaissance and preparing to give birth to the French Revolution, chocolate spread to the rest of Europe.

There are several theories concerning its introduction into France. The first claims that Anne of Austria (1601–66), daughter of Philip III, was the promoter when she married Louis XIII (1601–43), son of Maria de' Medici (1575–1642). Another says that it was Spanish monks who brought chocolate as a gift to their French brethren. Finally, there is the hypothesis based on the account of René Moreau (1578–1656), a renowned Parisian physician, which suggests that chocolate was first brought to France for medicinal purposes. He reported on the interest expressed by Cardinal Alphonse de Richelieu (1582–1653) in the therapeutic properties of cacao before 1642.

There is no denying that there was some confusion in the courts about the therapeutic benefits and purported shortcomings of chocolate (a substance that facilitated poisoning, promoted weight gain, aroused sexual desire, etc.). This ambiguity persisted until, in 1684, the Parisian physician Joseph Bachot presented his thesis on the drink, indulgently expressing that it was 'an invention so noble as to deserve to be the food of the gods, surpassing nectar and ambrosia'.

The invention of the chocolatier is attributed to the French, presumably towards the end of the 17th century or during the 18th:

> The roots of particular piece of equipment can be traced back to Mexico. It will be recalled that in pre-Spanish times, the native Mesoamericans prepared their chocolate by pouring the drink from one vessel to another. It will also be emembered that the 16th-century Spanish colonists introduced the *molinillo* … They discovered that a better result could be obtained by covering the pot with a wooden lid, with a hole in the middle for the *molinillo* handle. Such chocolate-frothers, made of lathe-turned wood, can be bought in many rural markets in today's Mexico, and were probably common in the Iberian peninsula. … But the French innovation seems to have been to fix a straight wooden handle to the metal pot at right angles to the spout; this handle was usually unscrewed clockwise so that it would remain tight while pouring from the pot in a counter-clockwise motion.. At the top was a hinged lid, with a central hole under the swiveling (or hinged) finial to take the handle of the *moussoir* ("froth-maker"), as they called the *molinillo*.
> Sophie D. Coe, Michael D. Coe, *The True History of Chocolate*, London, 1996, pp. 160–1.

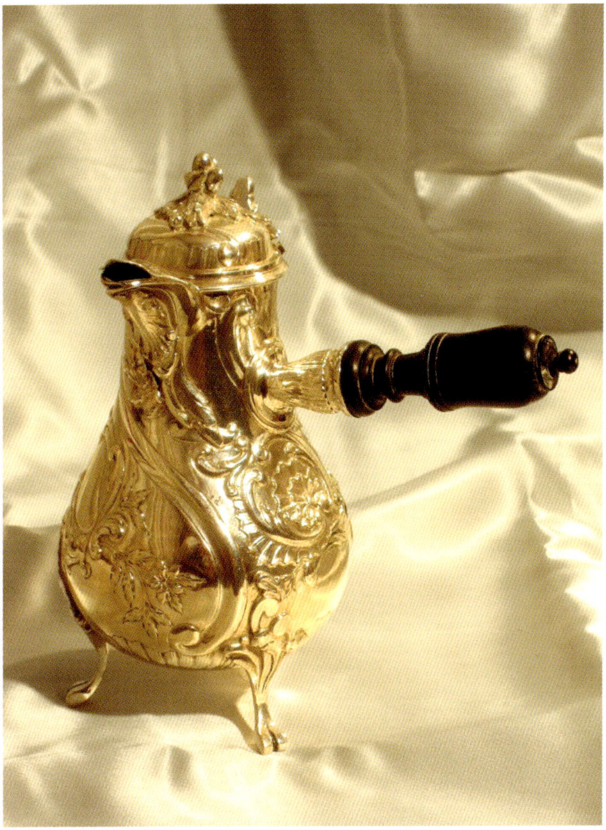

Chocolate became enormously popular when another princess, Maria Theresa of Spain, who was an enthusiastic chocolate consumer, married Louis XIV in 1660. It was then that chocolate officially conquered France. On the other hand, while in Spain and France the use of the drink remained a luxury reserved for the aristocracy, after Cromwell's troops wrested Jamaica from the Spanish in 1655, the island became the main source of cacao imports for the British. England was already familiar with the three big alkaloid beverages: tea, coffee and chocolate, but at this time the country began to bring in large quantities of cacao beans, similar to the way the Dutch West India Company began to bring in cacao beans from its Caribbean colonies and from Venezuela to Amsterdam, the capital of all traffic in goods. This made the 'drink of the gods' accessible to all social classes on the European continent who could afford to buy it. [R.V.]

With the exponential increase in cacao production in the American plantations of the colonies, the dizzying increase in the importation of cacao beans into Europe, the subsequent spread throughout the European continent and the overcoming of elitist barriers, the true great metamorphosis of the divine foodstuff began.

In the late 17th century, England introduced two innovations. First, treats were made with cocoa, eggs, sugar, spices and milk. Later, Sir Hans Sloane (1660–1753), president of the Royal College of Physicians, upon his return from a trip to Jamaica in 1689, had the idea of replacing the water used to dilute the chocolate with milk, giving the product an even more pleasant taste. Yet again it was he who established a cause-and-effect link between the regular consumption of chocolate by Jamaicans and their remarkable state of health, in his book *A Voyage to the Islands Madera, Barbados, Nieves, S. Christophers and Jamaica…* (1707–25).

The Swedish naturalist Carl von Linnè (Linnaeus; 1707–78) changed the scientific name of cacao to *Theobroma cacao* ('Theo', god and 'broma', food) or 'food of the gods' in 1753. In 1759, Joseph Fry (1728–87) patented a modified version of Watt's steam engine and used it to grind cacao beans, marking the beginning of the mechanisation of chocolate production. The industrial revolution would begin the following year. Although the process of democratisation of cacao was enthusiastically welcomed by the public, particularly by the nascent working class, there was a decline in the prestige of the product, relegated to the status of a children's breakfast item and replaced by other beverages such as tea and coffee, considered more immediate and accessible sources of gratification. A turning point in the cocoa industry was to be entrepreneurs with extraordinary inventive ingenuity, thanks to whom chocolate underwent an accelerated metamorphosis, even changing its consistency from liquid to solid:

Chocolate powder. In Amsterdam, in 1828, Casparus Van Houten (1770–1858) patented a hydraulic cocoa press that separated the mass from the cocoa butter. An inexpensive process to extract the fat from roasted cacao beans, resulting in two distinct products: a solid cocoa cake that was reduced to a soluble powder and cocoa butter. Casparus's son, Coenraad Van Houten (1801–87), a chemist, further perfected cocoa powder, inventing the 'dutching' process ('Dutch process'), which consists of treating cocoa with an alkaline substance, such as sodium bicarbonate, to increase the pH. The result is a cocoa powder with a much less bitter taste than untreated cocoa.

Chocolate bars. In 1847, the Fry and Sons Company of Bristol created the first chocolate bar, moulded by removing excess cocoa butter and then slowly adding it back to the melted chocolate.

Milk chocolate. In 1875 in Switzerland, Henri Nestlé (1814–90), the inventor of condensed milk, worked with Daniel Peter (1836–1919) to create what would become the first milk chocolate bar.

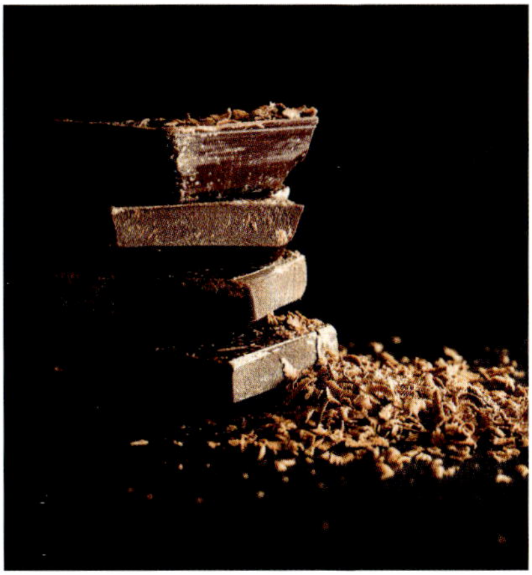

Dark chocolate.

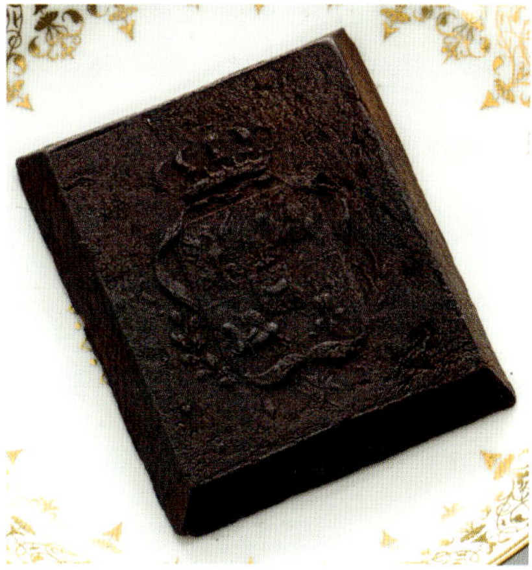

An 18th century chocolate bar.

Dark chocolate. In 1879, the Swiss Rodolphe Lindt (1855–1909) invented the 'conching' process, in which stone grinders were used on the cocoa paste, which was placed in shell-shaped containers, for a long and meticulous procedure involving intense stirring, shaking and aeration of the heated cocoa mass to remove unwanted acidity and bitterness (this process is not unlike evaporation). The 'conching' time also helps to dissolve any unwanted lumps, creating a velvety, silky chocolate texture. The chocolate produced by this method is the so-called dark chocolate.

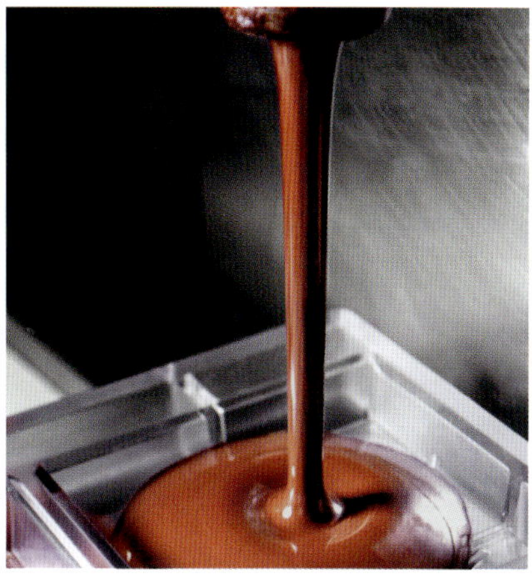

Dark chocolate.

All these innovations laid the foundations for a technological and industrial development in the cocoa sector beyond all expectations, generating a significant socio-cultural impact in the years to come, which was universal in scope but with ambiguous and conflicting implications. On the one hand the wealth and affluence of consumer societies and on the other hand the poverty, slavery and environmental unsustainability of those who produce the consumer good. [R.V.]

6. SLAVERY

> All they have done for the past forty years, and continue to do today, is torment, kill, harass and destroy them with extraordinary acts of cruelty … such that of the three million souls we have seen on the island of Hispaniola, not more than two hundred remain today.
> Fr. Bartolomé de Las Casas, *Brevísima relación de la destrucción de las Indias*, Seville, Sebastian Trujillo, 1552, fol. aVv.

The *cacaotales* (cacao plantations) of Chontalpa, in Tabasco, and Soconusco, in Chiapas, were the cacao havens of America. After the conquest, it was the colonists (the Spanish, the *criollos* and the privileged *mestizos*) of Nueva España who cultivated cacao in the region, exploiting the labour of the enslaved indigenous peoples in the *encomiendas*. It was Columbus who had 'sensed' that the only way to profit from early discoveries was to enslave the natives. According to the custom of the time, any prisoner of war who was not a Christian could be enslaved.

During the 16th century, Tabasco and Chiapas ceased to be the great cacao producers they had been before the conquest, followed by other regions of Nueva España. This was caused by the high mortality rate of the indigenous people in Mexico and America, which was the result of enslavement and epidemics and considered the worst demographic catastrophe ever known on the planet.

In 1514, the Dominican friar Bartolomé de Las Casas, a converted colonist and slave trader, as well as a theologian, bishop of Chiapas and fierce defender of the indigenous peoples, challenged the rights of the *conquistadores* and the *encomenderos*, denouncing the extent of an ethnocide carried out by colonial policy in La Española and Tierra Firme, the coastal strip of present-day Venezuela.

The numbers given by de Las Casas, according to historian and essayist Tzvetan Todorov (1939–2017) in his book *La conquête de l'Amérique. La question de l'autre* (Paris, 1982), are of the 'same order of magnitude' as those established today: 'it can be assumed that in 1500 the population of the globe was of the order of 400 million, 80 million of whom resided in America. By the mid-16th century, only 10 million of these 80 million remained. If we focus only on Mexico, on the eve of the conquest the population was around 25 million; by 1600 it was reduced to 1 million' (p. 138).

The enslavement of the indigenous peoples became a politico-religious problem, as they later became royal subjects, converted to the Catholic religion and therefore no longer enemies. Then endless arguments ensued, trying to prove whether or not they were human beings. The most famous was the 'Valladolid debate', a debate that approached the subject from all points of view: philosophical, theological and juridical, in the College of St Gregory in Valladolid. It started in 1550 and ended in 1551, without reaching an agreement or a judgement. It was, however, the first time in history that a topic was debated that has

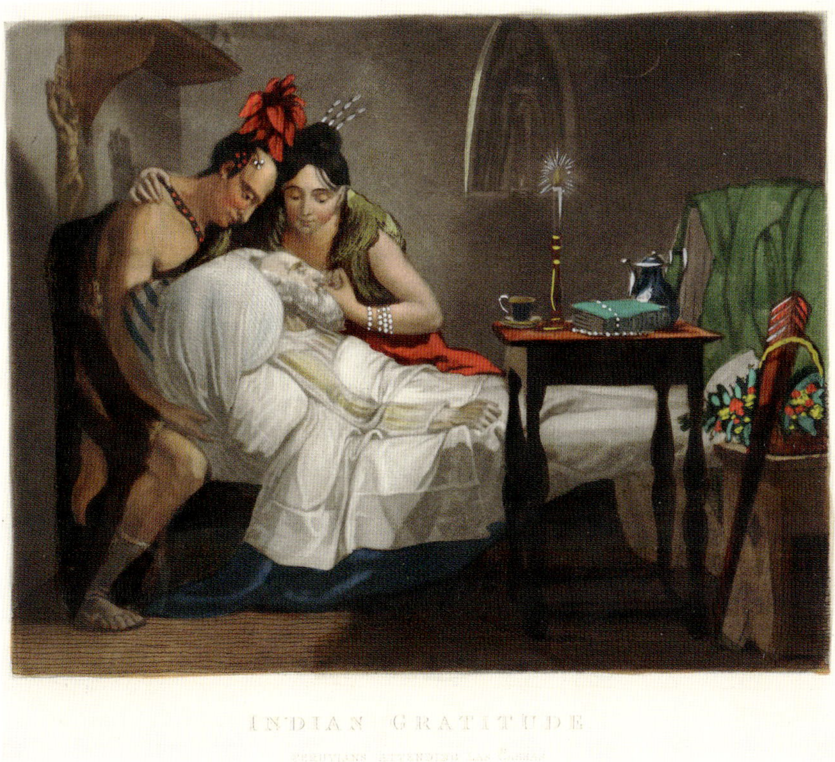

Unknown author, *Indian Gratitude*, *c.* 1840, engraving on paper, 18.5 × 22 cm, London, British Museum.

endured over the centuries and gradually emerged in modern political constitutions: all people are equal and have the same rights, regardless of their nationality, gender, race, religion or any other circumstance. [R.V.]

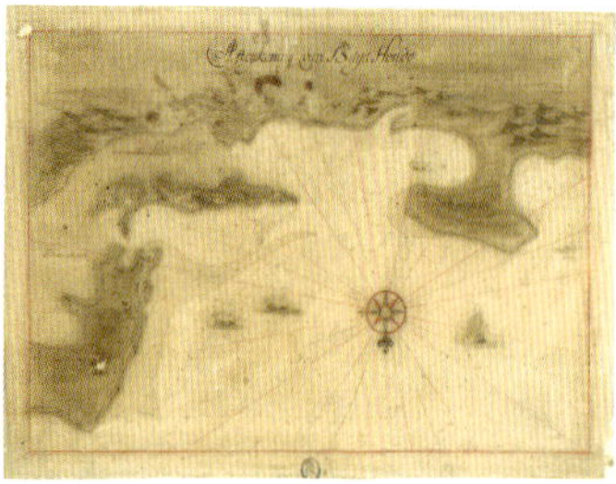

Attributed to Johannes Vingboons and workshop

View of Baya Hondo (known also as Golfo de Venezuela)

Amsterdam, *c.* 1665
Line-and-wash drawing on canvas-backed paper, 60 × 47.2 cm
Florence, Biblioteca Medicea Laurenziana, Carte di Castello, 35

Venezuela was discovered by Christopher Columbus on his third voyage, but was first explored in 1499 by Alonso de Ojeda, Amerigo Vespucci and Juan de la Cosa. It is the country that, throughout the 17th and 18th centuries, had established itself as the main exporter of cocoa to Europe. The most important port for shipping cocoa from the New World to the Old was the bay of La Guaira. The precious cocoa, also known as *caracas*, belonged to the 'criollo' variety. It was appreciated for its quality, although not as much as that of Soconusco or Chontalpa. Along with the 'criollo' variety, the 'trinitario' variety was also exported. During this period, King Philip V of Spain (1683–1746) gave the newly founded Guipuzcoana Company of Caracas the privilege of supervising trade between Spain and Venezuela in 1728. This mercantile company would play a key role in the development of crops such as coffee, cocoa and indigo, which were previously virtually non-existent in the region, and soon became Spain's largest monopolistic trading company. The success was mainly attributable to the cacao trade, which experienced exponential growth as it spread across the Old Continent. In Europe during the Baroque era, a veritable craze for chocolate consumption was unleashed, involving practically the entire continent, leading to a doubling of demand. Millions of workers were needed to grow, harvest and process both sugar and cocoa.

On 29 May 1537, Pope Paul III Farnese issued the bull *Sublima Deus*, excommunicating any Christian who enslaved an indigenous person. Since the papal bull did not include individuals of African descent, slave traders began to consider Africa as a source of labour. When the native slaves died, they were replaced by African slaves.

In theory, the cacao trade from Venezuela, as well as from other regions of Latin America, was strictly regulated by the Spanish Crown; however, in practice, merchants and farmers on the coast were active smugglers, so an increasing part of their trade ended up in the hands of privateers sailing in the Caribbean Sea or the Pacific Ocean. Almost all were involved in the slave trade. [R.V.]

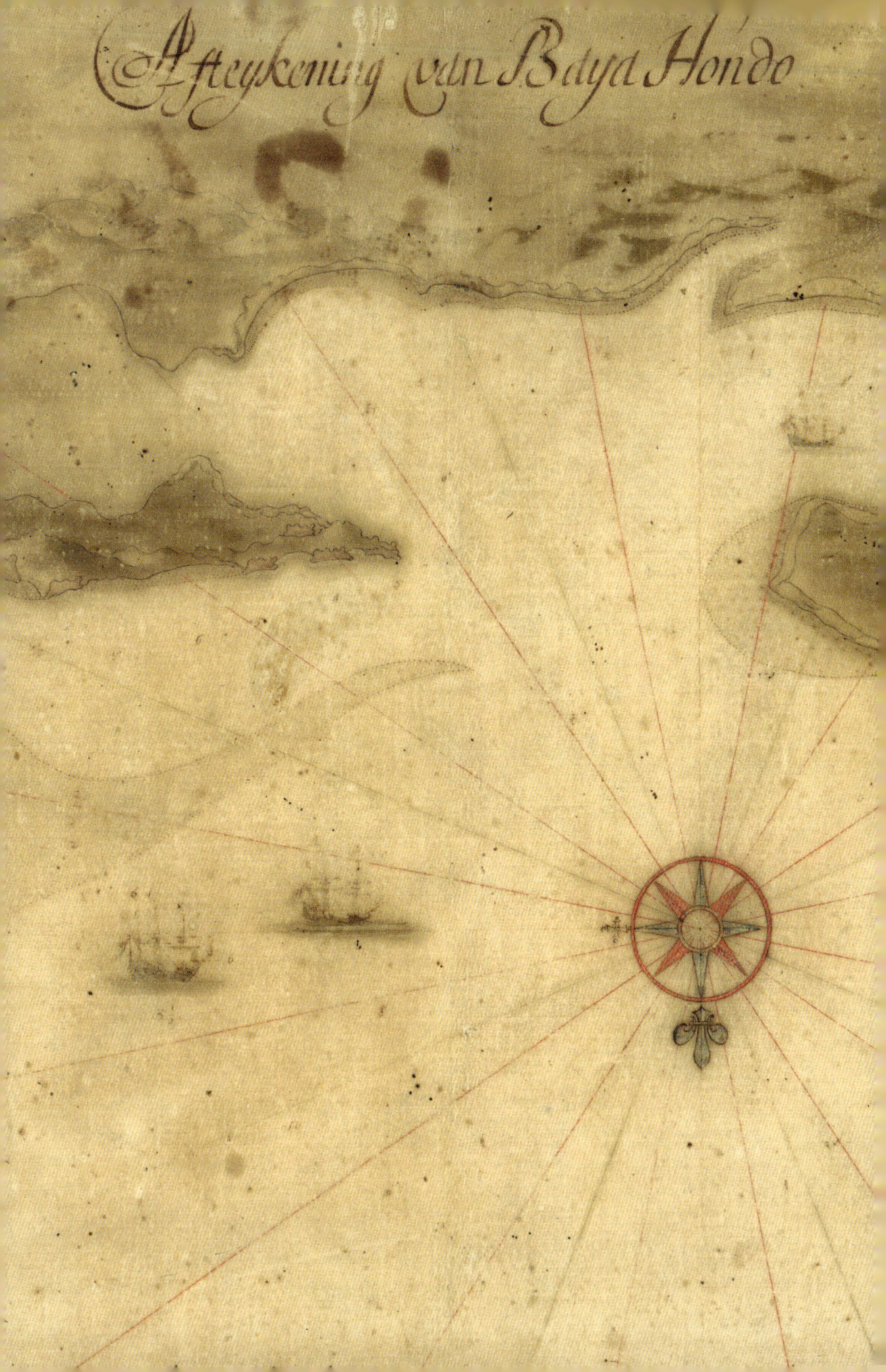
Afteykening van Baya Hondo

1619. The Portuguese slave ship San Juan Bautista, bound for the port of Veracruz, was intercepted by two British privateer ships, the White Lion and the Treasurer, which seized their cargo of Angolan prisoners. It is believed that most of the transported slaves had been imprisoned during an ongoing war between Portugal and the kingdom of Ndongo (Angola was a Portuguese colony). Both privateering ships were owned by a powerful English nobleman; Robert Rich, who was hostile to the Spanish and Catholics and profited from obstructing Spanish maritime traffic in the Caribbean. The White Lion, sailing under the flag of a Dutch port known to be half-pirate, later arrived at Point Comfort (Virginia) at the end of August 1619, followed four days later by the Treasurer. There, the governor and the head of commerce bought the '20 and some blacks' on board in exchange for 'victuals'. That is to say, they traded food for slaves.

This fact represents a milestone in the long history of slavery in the European colonies and in terms of the early stages of what would become the institution of slavery in America. The event marked the beginning of the trafficking of African men and women known as the 'Atlantic trade' or 'triangular trade'.

The Atlantic trade involved an intricate three-way system: slave ships transported finished goods from the European nations, such as clothing and weapons, to slave collection centres in Africa. In these centres, the European goods were exchanged for slaves, who were then embarked on ships bound for the New World. Once the slaves arrived at their destination, they were sold and sent to work on plantations. Slavers bought the products of these colonial plantations, then returned to their homeland and sold them on the European market. Until the 19th century, the Dutch, British, Portuguese and French exploited the colonies for large-scale production, facilitated by the Atlantic trade.

> The history of colonial products – sugar, cocoa, coffee – is embittered by the clashing chains of millions of human beings deported as slaves from Africa to the Americas. Between the 16th and 19th centuries, some 12 million Africans disembarked on the other side of the Atlantic.
> Stefano Carnazzi, 'Schiavitù, ieri e oggi. Una parola antica, un'omertà moderna', *LifeGate Daily*, 1 December, 2016.

Slavery is an ancient institution, known to most cultures and sanctioned by classical philosophy and official religions, but slavery on New World plantations was different, partly because of the strong racial component and partly because it was subject to the powerful new forces and pace of consumer capitalism and accumulation. The impact of the Atlantic slave trade on the global economy was decisive and affected all sectors of the European economy, even in countries that owned neither colonies nor slaves. The price in human lives and suffering was incalculable and terrible. 'Its damaging legacy still has repercussions today in most societies in Africa, Europe and the New World' (Gabriel

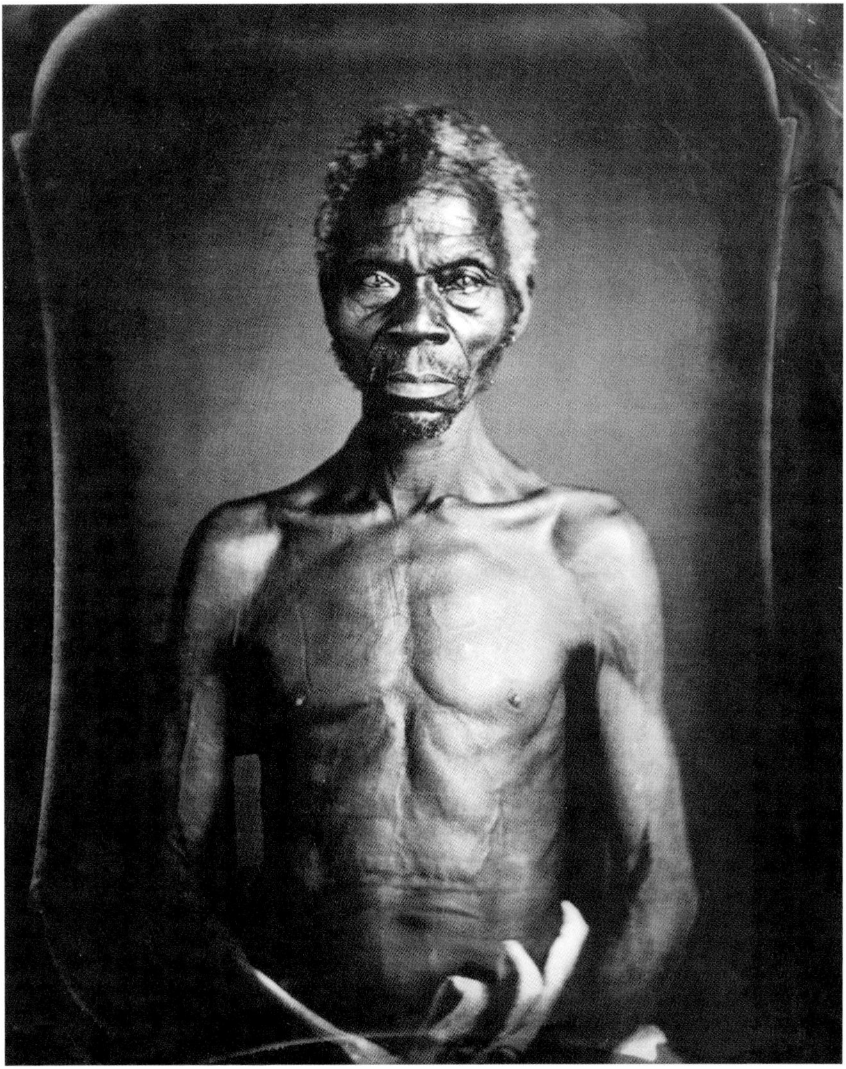

Louis Agassiz, *Renty, an African Slave*, Brazil, 1865.

Paquette, *Human trafficking across the Atlantic*, 'Historical National Geographic', 5 January 2022). [R.V.]

A 'Confederate States of America' 100-dollar banknote dated 22 December 1862, issued during the American Civil War (1861–5). Three African slaves are working on a cotton plantation in the centre, with politician John C. Calhoun on the left and Columbia, the female personification of the United States, on the right. More than 670,000 of these banknotes were issued between August 1862 and January 1863.

In the thirteen British colonies, the indigenous people were initially captured as prisoners of war. Later, the practice of selling them to Caribbean plantations and freeing up land for the expansion of British colonies began. This system, which was supported and encouraged by the divided indigenous tribes themselves, persisted throughout the colonial era until 1750. However, after the American War of Independence (1775–83), the Atlantic trade in African slaves became more profitable. [R.V.]

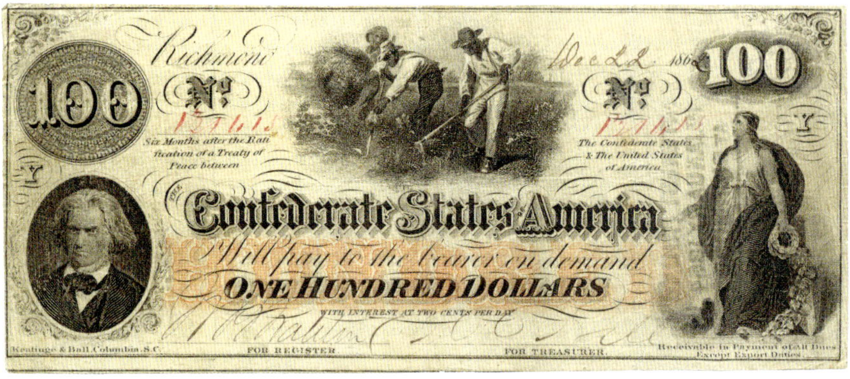

'Confederate States of America' 100-dollar banknote, 19th century.

Attributed to Johannes Vingboons and workshop
Elmina Castle seen from the mountain of St Iago

Amsterdam, before 1667
Line-and-wash drawing on canvas-backed paper, 60 × 47.2 cm
Florence, Biblioteca Medicea Laurenziana Carte di Castello, 57

The town of Elmina, Ghana, was one of the first to catch the attention of Europeans in the late 15th century because there was a gold mine in the area. Its name may derive from the Portuguese 'a mina' (the mine), but also from the Arabic 'al-mina' (the port). It was at that time that the Portuguese built what is considered the oldest European building in all of Africa's tropical territories. This construction combined European and African materials, with its walls and towers built with stone from a local quarry and bricks brought from Portugal. The Portuguese built the castle with the intention of using it as a commercial hub for the export of resources such as ivory and gold. Later, the Dutch West India Company seized the castle in 1637, expanding the slave trade and converting the fortress into a place where local slavers transported prisoners ready to be bought by European merchants. The Dutch sold it to the British in 1872.

From the 17th century, Elmina, with its imposing castle, became the main slave market, one of about 40 forts and castles built for the purpose of 'storing' enslaved Africans.

The slaves were captured in their homelands and then crammed into the underground dungeons of castles waiting to be bought or bartered for European goods. The captured slaves spent between six and twelve weeks in the castles, where more than 1,000 human beings could be packed. The main gates were called 'Gates of No Return', through which

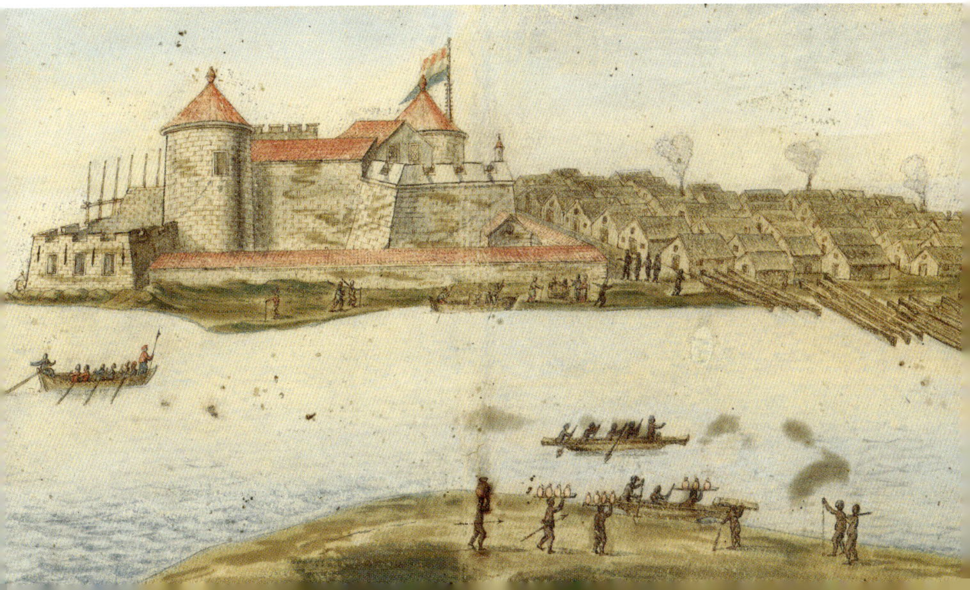

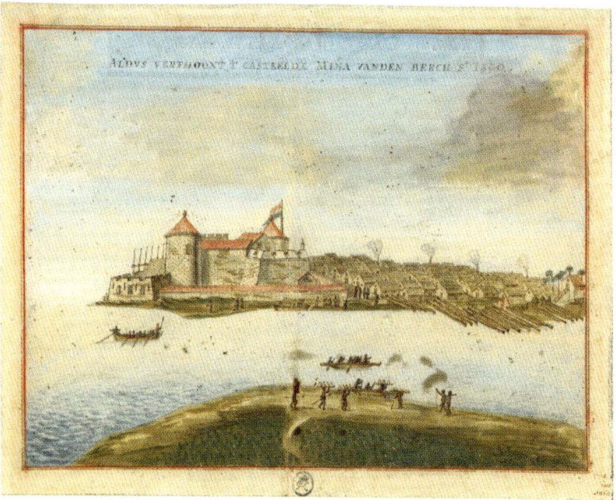

thousands of slaves passed, never to return to their homes. For the slaves, this was often the last glimpse of their homeland before being transported on ships across the Atlantic.

The ships of the Atlantic trade were not simply trading vessels. They were also factory prisons, or floating prisons, in which prisoners were transported subject to violent discipline. During the journey, the slaves were chained, handcuffed and shackled, naked and malnourished. It was a life-threatening voyage also because of the poor hygienic conditions on the ships. It is estimated that between 300 and 450 slaves could be crammed onto a single ship and treated like animals on a voyage that would take at least 100 days. Those who tried to resist were beaten and tortured; some were even left in the hot sun to die of hunger and thirst. The sick were thrown into the sea and women were subjected to violence.

The slave ports of the Spanish colonies were San Juan in Puerto Rico, Santo Domingo, Havana, Cartagena de Indias and Veracruz. At the same time, the territories occupied by the Netherlands, the British and the French became major centres of slave distribution: Barbados, Jamaica, Dominica, Martinique, Guadeloupe, Curaçao, Saint Martin, Saint Croix, Saint John and Saint Thomas. In addition to the ports of the Thirteen Anglo-Saxon North American Colonies. [R.V.]

ESCLAVA

Proof of purchase of a slave

Collezione Roberto Caraceni

ADMINISTRACION GENERAL DE ALCAVALAS. LIMA.
Dona Maria Negron has paid into the general treasury of this Royal Customs twenty-seven […] for the Alcavala (sales tax) at 6% of the debt of four hundred and sixty, for which she sold to Dona Maria Ignacia Talavera a slave named Maria Mercedes. It is hereby advised that without this document the scribes will not be able to draw up the notarial deeds nor issue any receipt for the contracts to the parties concerned. This was noted in the corresponding account at this General Accounting Office on 16 June 16-781 [16 June 1781].

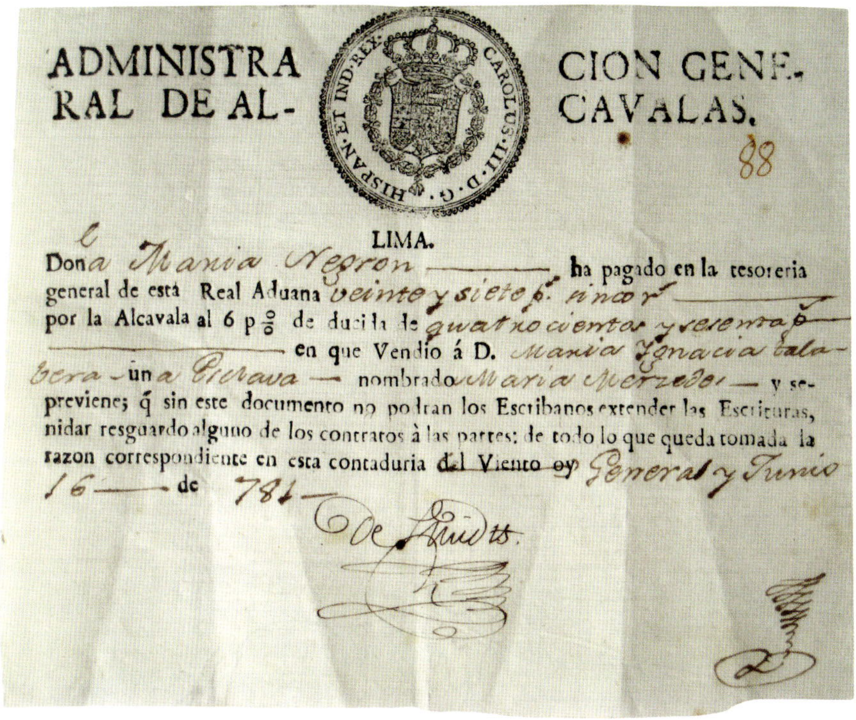
ADMINISTRA CION GENE-
RAL DE AL- CAVALAS.

CAROLUS·III·D·G·HISPAN·ET·IND·REX·

88

LIMA.

Dona Maria Negron ———— ha pagado en la tesoreria general de està Real Aduana veinte y siete p. cinco r ——— por la Alcavala al 6 p% de dicha de quatrocientos y sesenta p ——— en que Vendio á D. Maria Ygnacia Talavera una Esclava — nombrada Maria Merzedes — y se previene; q sin este documento no podran los Escribanos extender las Escrituras, ni dar resguardo alguno de los contratos à las partes: de todo lo que queda tomada la razon correspondiente en esta contaduria ~~del Viento oy~~ General y Junio 16 — de 781 —

ABOLISH

In the 18th century, the term 'right' in relation to the subject was in common use. Enlightenment culture made a fundamental contribution to the dissemination of these ideas. Among the intellectuals of England, France, Germany and Italy, philosophical schools were formed that examined the very foundations of society, questioning the legitimacy of the Church, the nobility and even the monarchy. Predictably, these divisions were reflected in food and drink: chocolate was considered southern, Catholic and aristocratic, while coffee was northern, Protestant and middle-class. The aristocrats drank chocolate, while the middle classes preferred coffee. The working class consumed cheap wine in the south and beer in the north.

The intellectuals of the Enlightenment were advocates of man's rights to a free and prosperous life. The first to outline a plan to abolish slavery was Marie-Jean-Antoine-Nicolas de Caritat, Marquis of Condorcet, a leading exponent of the Enlightenment. This proposal was published in his essay *Réflexions sur l'esclaveage des nègres*, in 1781. This was followed in 1787 by the founding of the Society for Effecting the Abolition of the Slave Trade by Thomas Clarkson (1760–1846) and other abolitionists, to campaign against slavery worldwide. In France, following Clarkson's example, Jacques-Pierre Brissot de Warville (1754–93) founded his own abolitionist club in February 1788, the Société des amis des Noirs.

After centuries during which European powers battled over the lucrative slave trade, anti-slavery demands in the 19th century, triggered by the French Revolution and slave rebellions in the colonies, led to the first decrees for the abolition of slavery. Following some political efforts, slavery was effectively abolished in the British territories in 1833 with the Slavery Abolition Act, in Denmark in 1846, in France in 1848, in the Netherlands in 1863, in the United States in 1865 and in Portugal in 1869.

The last country to abolish the practice of slavery was Brazil, in 1888. [R.V.]

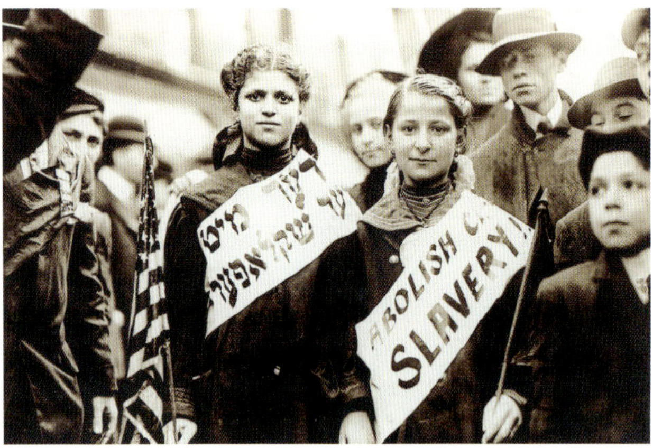

Bain News Service Photograph Collection, *Abolish Child Slavery*, USA, 1909.

GLOBAL

Southafrica 1900

United Kingdom, 20th century
Metal
Associazione Museo del Cioccolato private collection

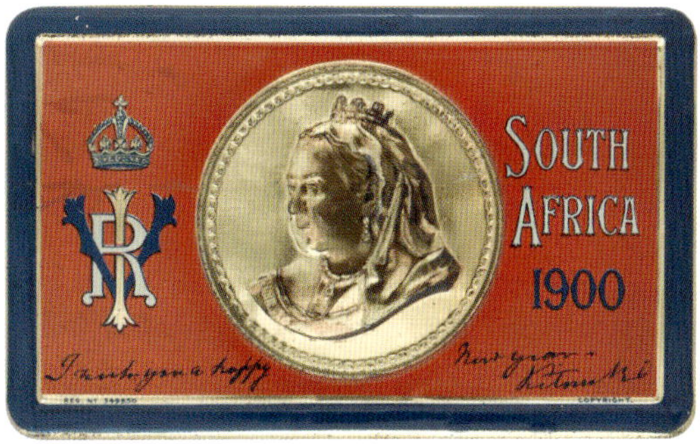

At the same time that the abolitionist winds began to blow, the cacao tree from Brazil was being transplanted to Africa. In 1824, the 'forastero' variety of *Theobroma cacao* was introduced into the Portuguese colony of São Tomé (Equatorial Guinea), from the lucrative Jesuit plantations in Brazil. The Gold Coast (Ghana), Nigeria and the Ivory Coast followed. During the 19th century, the Germans began growing it in Cameroon, the British introduced it into Sri Lanka, and the Dutch brought it to Java and Sumatra. By the end of the 19th century, European colonialist states had established cacao tree plantations practically along the entire equatorial belt. Previously, the traditional supplies of 'criollo' and 'trinitario' cacao beans from Venezuela had already been replaced by exports of 'forastero' cacao from Guayaquil, in present-day Ecuador.

In the meantime, famous chocolatiers and 'literary cafés' were established as centres of socio-cultural influence, such as Café Florian in Venice and Antico Caffè Greco in Rome; Le Procope and El Flore in Paris; Café Central in Vienna; Café Gijón and Café Comercial in Madrid and Café Novelty in Salamanca; The Chocolate House in London; the A Brasileira in Lisbon; Café Slavia in Prague; The White Horse in New York; La Biela in Buenos Aires; the Casa de los Azulejos and La Ópera in Mexico City.

The advent of new technologies, the increase of land for cacao cultivation and even soldiers contributed to the universal spread of chocolate consumption. It was 1900 when Queen Victoria sent chocolate on New Year's Eve to every British soldier fighting the Boer War in South Africa, to lift their spirits. From then on, chocolate became an integral part of the military rations of first the British and then the US armies. By the end of the Second World War, the popularity of chocolate had spread to almost the entire globe, with the exception of the Orient and Asia, where its consumption remained limited mainly to the Philippines. [R.V.]

> When the great transformation from liquid to solid took place, the technological breakthroughs that made it possible were pretty much confined to the Protestant and thoroughly capitalistic nations of northern and central Europe; the Catholic countries of the Mediterranean, which had played such a leading role in chocolate's history in preceding centuries, were now largely backwaters as far as such innovations went. In isolated pockets within southern Europe there were a few picturesque survivals of ancient ways of making chocolate.
> Sophie D. Coe, Michael D. Coe, *The True History of Chocolate*, London, 1996, p. 236.

At the beginning of the 20th century in Latin America, Italy, Spain and the South of France, there were still chocolatiers, travelling from house to house with their stone *metate* to prepare chocolate following the ancient method of cold processing of the cacao.

Apart from these exceptions, after the second half of the 20th century chocolate had mainly become an industrialised product and was consumed mainly in solid form, with a vast choice of varieties of packaging and flavours. As is usual in large-scale food production, some were subject to adulteration by unscrupulous producers and traders, with vegetable oils, starches and/or earthy substances and other solids. Because it was affordable on the large mass market, it was regularly consumed within families belonging to the middle classes. However, it had two main target audiences: firstly, children and secondly, women. Chocolate manufacturers were able to associate certain characteristics of their products with the traits attributed to the 'nature' of the female and the child, through the visual allure of advertising. It was during this period that the culture of promoting lifestyles as consumer goods was created. The international reputation of branded chocolate products was also established. Among the most prominent are: in Italy, Gianduia (Turin), Baci (Perugia), Nutella (Alba); in the United Kingdom, the egg with a surprise (Bournville), After Eight and Smarties (York); Pralines (Paris and Belgium) and Truffles (Paris and London); in Switzerland, Nesquik (Vevey) and in the United States, Kisses (Hershey). The end of the 20th century was marked by the establishment of massive economic conglomerates in the food sector.

As we enter the 21st century, thanks to growing awareness of health and environmental issues, people are more aware of the importance of making good consumer choices. A branch of the world of chocolate is busy revolutionising the heavenly food yet again, with the intention of once more bringing the quality of the product to the forefront. The concept of cacao of 'origen único' and 'fino de aroma' was established, and expressions such as 'grand cru' chocolate were introduced. Today, artisanal chocolate is a combination of culture, ingenuity and science. The creations of the master chocolatiers are hailed as genuine works of art and delicacies of haute cuisine. This is in contrast to the vast industrial production of multinational corporations, with their oppressive economic and political power. [R.V.]

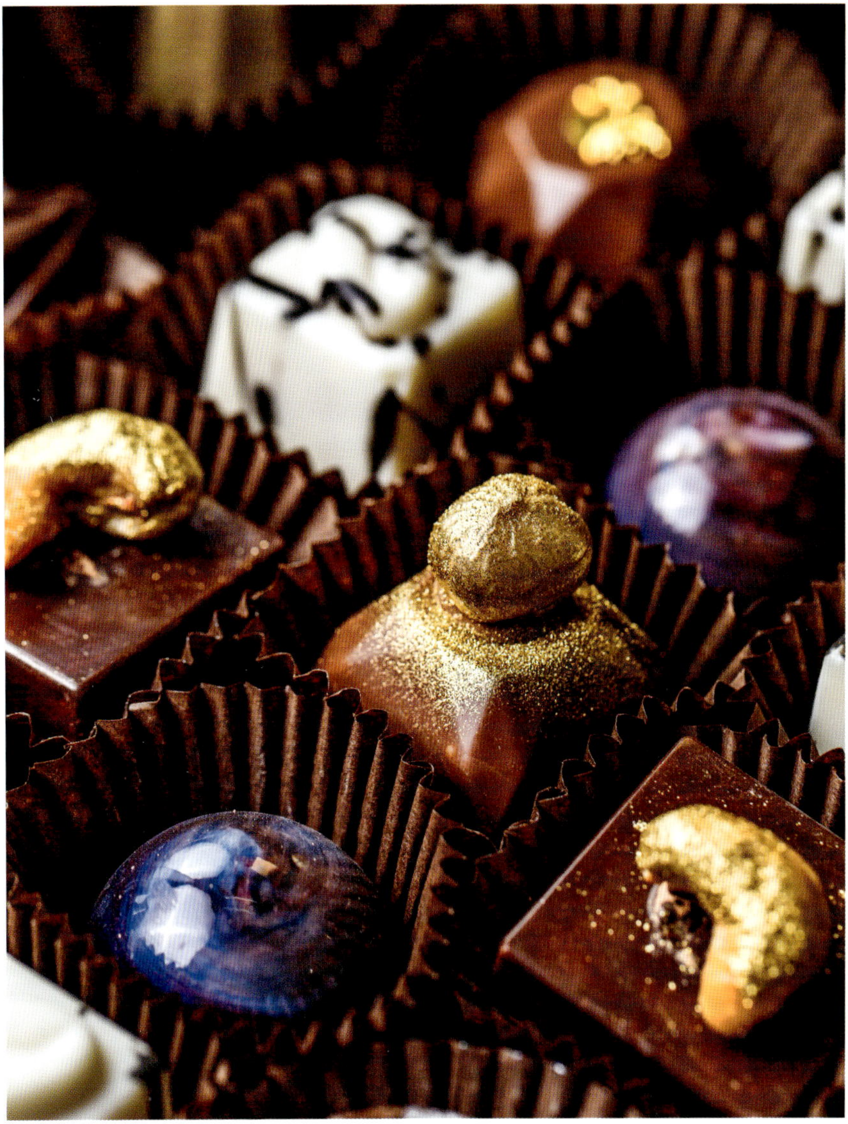

Chocolates.

FORASTERO

Cocoa and chocolate are part of a global market economy involving most countries in the world. *Theobroma cacao* 'criollo' is grown mainly in Mexico, Venezuela, Colombia, Peru and, recently, in Hawaii. *Theobroma cacao* 'forastero' is grown in Africa, Asia and South America. 'Trinitario' cacao is grown in the Caribbean and Asia. All three varieties are grown in Ecuador.

For economic reasons, the large chocolate manufacturers prefer 'forastero', which currently accounts for over 80% of production worldwide. Africa is its largest producer, and at the beginning of June 2023 it produced 3.73 million tonnes of cacao beans, which is 74.8% of global production. 60% of this figure is concentrated in Côte d'Ivoire and Ghana, and of the two, the country with the largest production is Côte d'Ivoire, with 2.10 million tonnes of cacao beans. In November 2023, cacao prices reached USD 4458 per tonne in New York, an increase of 27% compared to the initial values for the 2023/4 season. The global cacao and chocolate market is estimated to grow from a value of USD 68 billion in 2029, according to analysts at Fortune Business Insights.

Most chocolate production is carried out by machines in factories. The mechanised process is complex and laborious, similar to an assembly line: the cacao beans are mixed, roasted, cracked, fanned, ground, pressed, shelled, refined and tempered to become chocolate bars. Although the industry has made the production of chocolate faster and its purchase cheaper, it has not changed the way cacao is grown, which is mainly manual. Chocolate manufacturers must purchase cacao from independent farmers or cooperative groups that manually grow, harvest, ferment, dry and package the beans. [R.V.]

Map of the world distribution of the cacao tree.

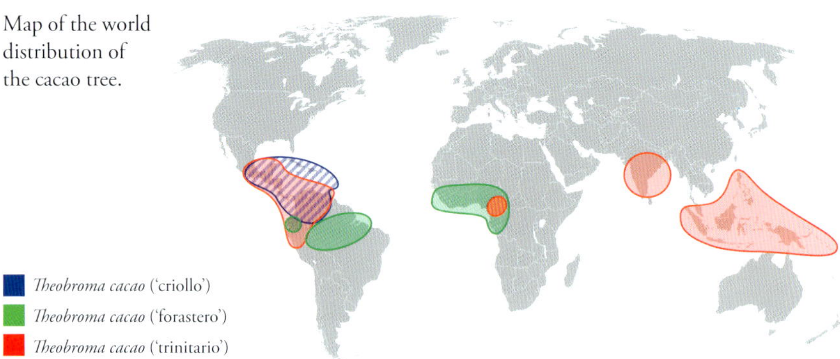

> Not even the little children would go near the fruit of the cacao tree. They were afraid of that yellow, sweet-seeded cacao, which made them prisoners of that life of dried meat and *jaca*. Cacao was the great lord, of whom even the *coronel* was afraid.
> Jorge Amado, *Cacau*, Rio de Janeiro, 1933 (1st edition).

Although today almost the entire production process is handled by machines, the cacao is harvested by hand so as not to damage the buds that produce the pods. The harvesting has to be carried out with extreme care, requiring human intervention, which, however, still implies, even in the middle of the 21st century, the exploitation of women, youths and especially, children. Equatorial African cacao (the 'forastero' variety) is the one that costs the big chocolate companies the least precisely because it involves enslaved African children in its supply chain. Exploiting them as labour helps to reduce the costs of the final product. They are whipped, beaten and treated like animals to collect the cacao beans, which are then made into chocolate sweets for more fortunate children in Europe and America.

> The largest chocolate companies in the world promote child slavery and the abuse of women in cacao farms and in their supply chains.
> International Rights Advocates (IRA), 2021.

A report by the University of Chicago shows that in 2019 as many as 790,000 children, including some as young as 5 years old, were employed on cacao plantations in Côte d'Ivoire. Scholars have also identified a similar situation in the neighbouring territories of Ghana.

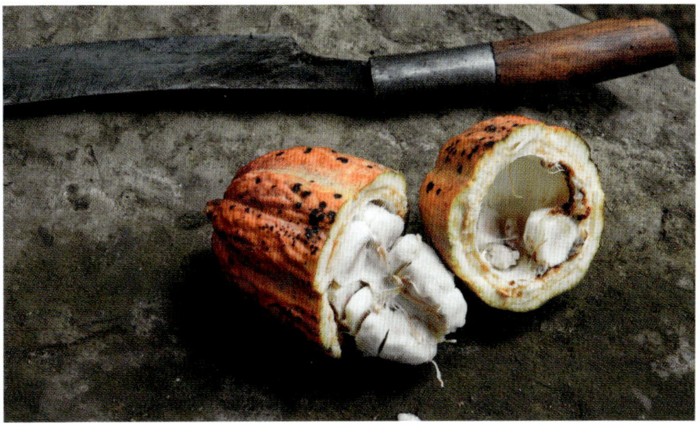

Cacao *cabossa*.

These non-children are forced to carry excessively heavy loads on their shoulders, which damage their spinal columns; use dangerous tools with which they often injure themselves, such as machetes; and breathe in all the chemicals used on the plantations on a daily basis. But the chocolate manufacturers continue to hide behind co-financing pilot projects that aim to change farmers' practices, without them having to change their own. Operations that smack of sordid social greenwashing.
Roberta Ragni, in *La Stampa*, 5 January 2012.

In 2001, a protocol to combat child slavery in cocoa production by 2005 called the Harkin-Engel Protocol was formalised with the endorsement of the International Labour Organisation; the governments of the United States and Côte d'Ivoire and the International Cocoa Organisation, and signed by the manufacturing associations. As pacts were then broken, the protocol was renewed with a promise to reduce child exploitation by 70% by 2025.

On 12 February 2021, seven multinational food industry companies were sued in a US federal court for alleged trafficking and forced child labour. The human rights group International Rights Advocates filed a class action lawsuit on behalf of eight victims of trafficking and child slavery on cacao plantations in Côte d'Ivoire. The lawsuit was filed on the basis of the US Trafficking Victims Protection Act (TVPA), which allows victims of trafficking and forced labour to sue companies that benefit from these practices. The IRA noted that these companies profit from trafficking and forced child labour, promoting 'bogus programmes that falsely claim to solve the problem'. [R.V.]

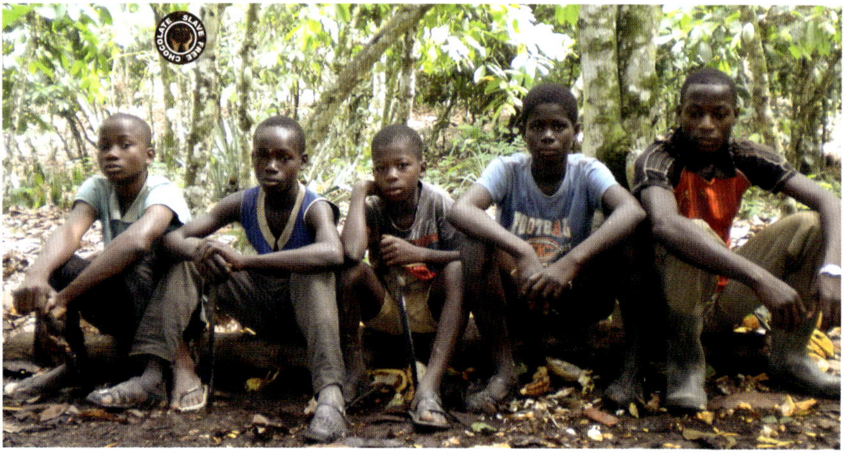

Children employed on a cacao plantation in Côte d'Ivoire.

In June 2022, the seven multinationals won, with the dismissal of the case by the US federal court, and the decision to block the trial.

The exploitation carried out by these global corporations does not only involve the violation of human rights. To meet the current demand for cacao beans for their chocolate factories, the major production areas are being deforested because of intensive farming, and cacao trees can be seen in regular rows in areas where there used to be tropical forest. Over the last sixty years, Côte d'Ivoire and Ghana have lost between 80% and 90% of their forests. It is estimated that about one third of this loss is attributable to the cocoa sector.

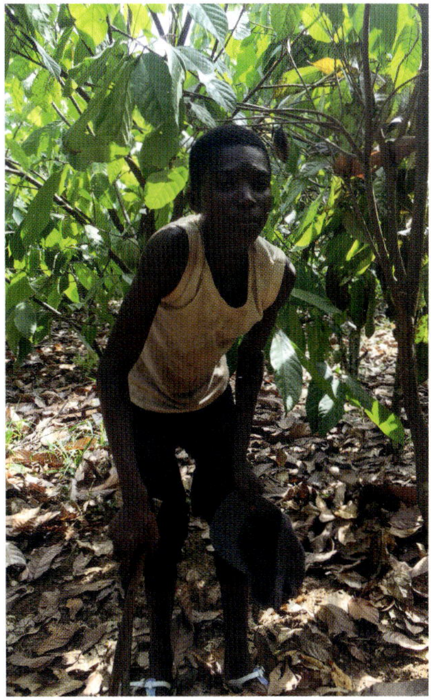

Cocoa-gathering child in Côte d'Ivoire.

The position of the multinationals on profits makes these scenarios even more grotesque. A new study by the organisation Oxfam, published in May 2023 on the occasion of World Fair Trade Day, revealed that the largest chocolate companies make huge profits but do not pay enough money for cacao producers to earn decent incomes.

> The current supply shortage, which has led to the current price increases, is mainly caused by unfavourable weather conditions in Côte d'Ivoire and Ghana. The high cost of key ingredients due to supply restrictions will inevitably lead to higher production costs for chocolate manufacturers who have indicated that increases in the cost of key ingredients will be passed on to consumers. For example, Mondelez has stated that it will increase the prices of some of its products as the company is under increasing pressure due to the costs of cacao and sugar. At the time of writing, reports indicate that some major international cocoa purchasing companies, including Barry Callebaut AG, Cargill Incorporated and Olam International Limited, are in a standoff with Ivorian government negotiators and are suspending cocoa purchases pending a price discount.
> International Cocoa Organization, Market review (November 2023).

The billionaire chocolate multinationals, instead of taking a reduced profit margin because of the increased cost of cocoa due to adverse weather conditions, are putting pressure on African farmers in negotiations for the purchase of cocoa, seeking to obtain discounts through coercion. It is also worrying that the International Cocoa Organisation defines the price of cocoa as 'high cost', set at around EUR 1.05/kg.

> … the priorities are profits and power … human rights are of instrumental value for propaganda purposes, nothing more.
> Noam Chomsky, *Year 501: The Conquest Continues*, London, 1993, p. 12.

Until multinationals are willing to review their business practices, no initiative aimed at combating child labour or slavery, promoting environmental protection or ensuring fair compensation for agriculture will ever achieve the desired sustainability goals. [R.V.]

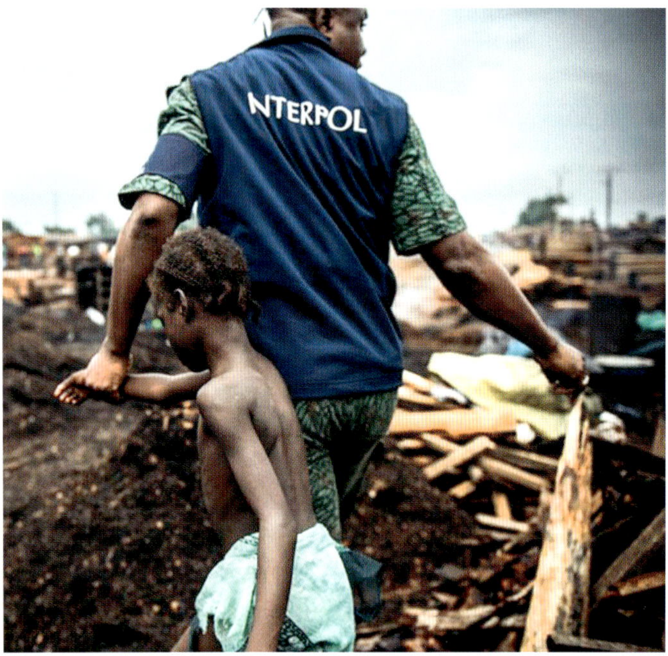

Interpol agent in operation against child exploitation.

Each one of us can act to help put an end to such criminal actions. In addition to educating ourselves, our family and friends to think critically, we need to buy and consume chocolate responsibly. We are called upon to make conscious choices, opting for products that do not involve human and/or environmental exploitation. 'Conscious indulgence' should become a widespread trend among consumers and establish itself as a compelling demand in a growing market.

As a first step towards awareness, the different critical aspects of sustainability in the world of cocoa should always be taken into account: cost, consumption, production, equity. Let's start with the low cost of chocolate, which also encourages its daily consumption. Chocolate 'shouldn't' be cheap: it wasn't in Mesoamerican times and it shouldn't be today. There is a reason why in the times of the Maya and Mexica it served as currency and was a food consumed only on special occasions: because of its intrinsic and production value. The cacao tree was not abundant in the tropical Mesoamerican forests and its production was limited by environmental factors, so it did not grow in large numbers. Each fruit produced an average of just twenty beans. That's why it was a precious commodity, a rare product of the highest value. However, the cacao tree is still a delicate organism and requires special conditions and a lot of care in order to grow and produce fruit. Its production is labour-intensive, and pure, authentic chocolate is of great value both nutritionally and in terms of flavour. It is only right to remunerate a product according to its value, and therefore it is imperative that the price of chocolate be in relation to the effort put into its production and to its intrinsic qualities. To do justice to its delicacy, we should return to consuming it only on special, significant, festive occasions, as is customary with champagne and bourbon.

As far as its cultivation is concerned, the cacao tree should not be forced into large-scale production. A return to natural production would stop the arbitrary deforestation associated with cacao plantations. Trade should not exist at the expense of the environment and human rights, while consumption should aim at quality, not quantity. This would make us responsibly aware.

One should strive for goodness in all of its aspects. [R.V.]

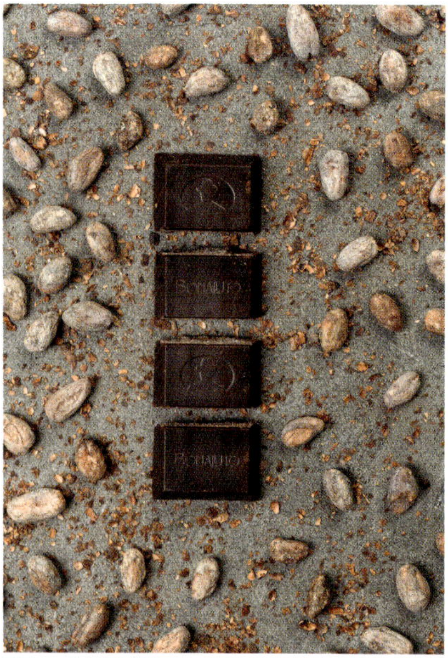

Chocolate bar.

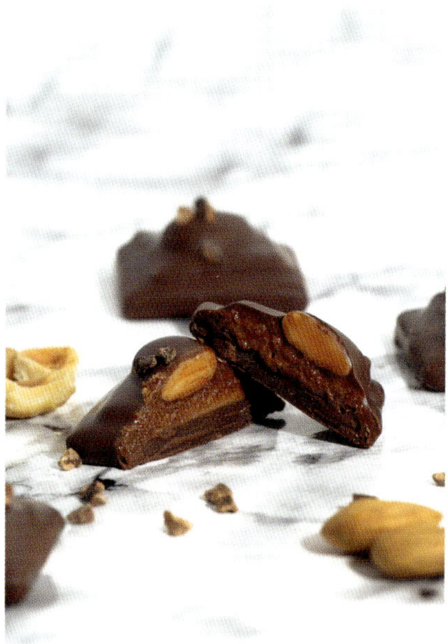

Dark chocolate 'bean to bar', 'Rivierotto' chocolates.

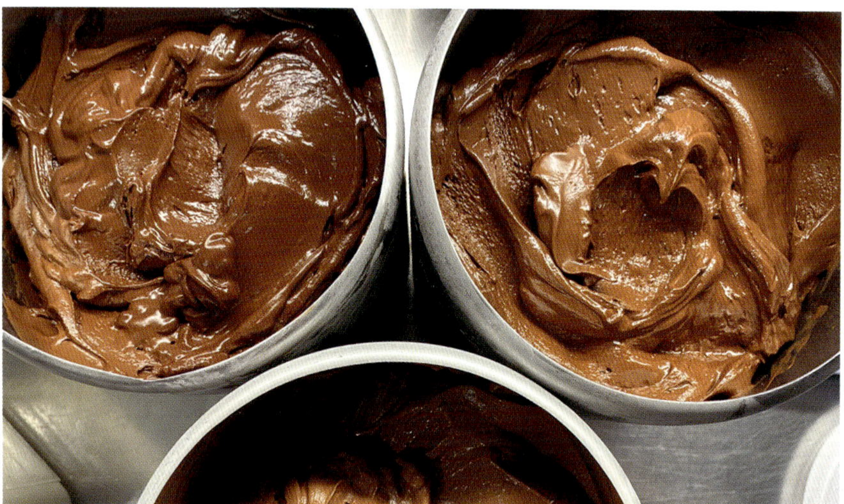

Chocolate ice cream.

BEAN TO BAR

One can truly speak of goodness, in the true sense of the word, when the world of chocolate refers to the concept of 'bean to bar' (Scharffen Berger Chocolate Maker, 1997).

'Bean to bar' refers to the artisanal production of chocolate. It is a movement that has gained ground in the world of cocoa in recent decades, as a response to the industrialisation of chocolate on behalf of multinationals that, on the one hand, do not do justice to the production chain and, on the other, offer a product that in most cases is of poor quality and, in some respects, bad for the health. Cacao is a noble fruit and its beans possess as many qualities of flavour and aroma as the *chocolate maker* is able to arouse, through the art of making chocolate from the cacao bean, similar to the art of making wine from grapes. In fact, cacao has varieties just like grapes, and each one requires different treatment by the chocolatier.

In general terms, the 'bean to bar' production process starts with the selection of the cacao bean at its place of origin, ensuring that the cultivation and harvesting take place according to strict standards of social and environmental sustainability, in small batches, so that each step in the production process can be handled with dedication and precision. The collaboration between the *chocolate maker* and the growers spans the entire production process. The purchase of cacao beans is direct, without intermediaries, directly benefiting the producer communities.

There are many small and medium-sized chocolate factories that guarantee an ethical origin, the highest quality and traceability of the cacao used, striving to maintain its organoleptic properties and to support economic processes that help improve the living conditions of the farmers and their communities. They are outstanding products in the world of cocoa that promote a return to the origins, a model of sustainable development.

We purchase from those who offer 'delicious kindness' with a taste of fairness and justice. [R.V.]

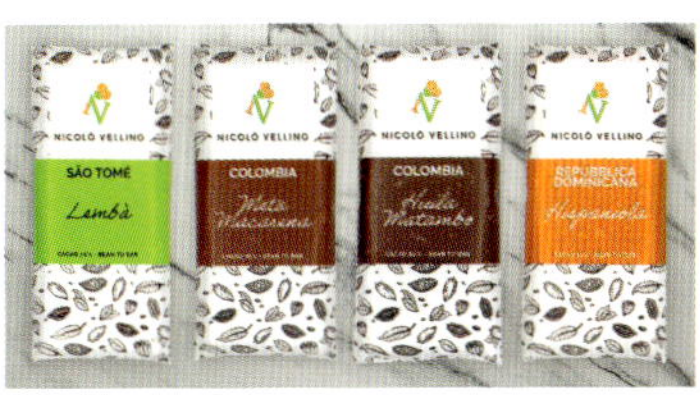

Chocolate 'bean to bar'.

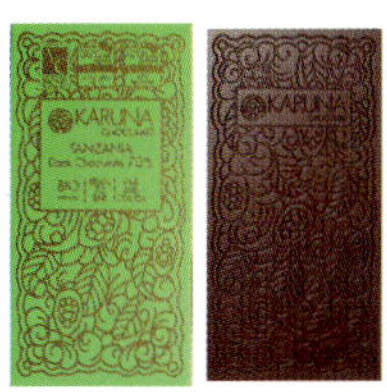

Chocolate 'bean to bar' from Tanzania.

7. MEMORY

MEMORY

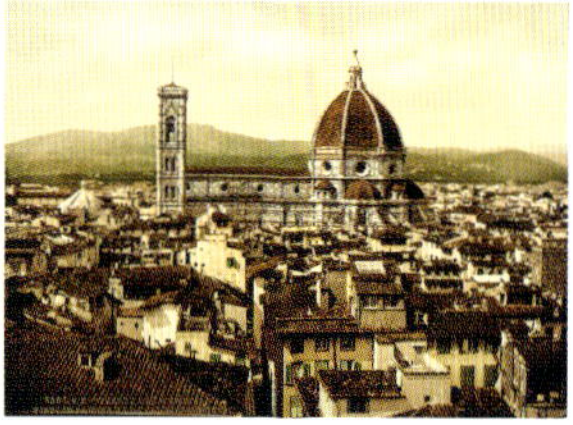

> If you do not want to repeat what has happened and can be repeated, fight with all your might against the logic of privilege!
> Laura Tussi, 'Riflessioni sulle minacce alla pace. Dialoghi con Moni Ovadia', *Faro di Roma*, 11 January 2024.

The market economy seems to have no values at its core other than satisfying the desires and profiting a privileged few. It is paradoxical that West Africa, now the world's leading cocoa producer, is the same region from which millions of men, women and children were snatched to work as slaves in the cacao plantations of the American continent. The process of enslavement of Africans, adopted as a European economic system until the 19th century, was the metastasis of a cancer that afflicted America in the 16th century. Just as today the trafficking and enslavement of children on cacao plantations is the metastasis of a cancer of multinational corporations. Historical developments and events are linked to a large extent to those of economic science and, vice versa, at the same time history is necessary in all three economic dimensions: theoretical, analytical and political. Similarly, economics and ethics both stem from a single discipline: Moral Philosophy. In the past, ethical discourse dominated the study of economics: Aristotle himself spoke of politics, economics and ethics as three inseparable subjects, which were to provide the means for the realisation of the human good. These are the scientific reasons why social justice and economic development go hand in hand: the lack of one inevitably compromises the other.

In order to set a limit in economic systems between what is possible and what is permissible, we need to draw on our historical memory, which is the key to awareness. It enables us to recognise the mistakes and injustices of the past to reduce the possibility of equally catastrophic effects in the present and in the future. Hence the importance of remembering the events that have shaped the consciousness of those who experienced them and, subsequently, of those who learnt about them, with a commitment to preserving these memories. This was the impetus behind Friar Bernardino de Sahagún's creation of the *Historia general de las cosas de Nueva España*:

> I have heard that the Viceroy and the Commissary were to send to Your Majesty these works divided into twelve books, in four volumes, on the ship that departed recently. If this is not the case, I humbly beg Your Majesty to order that I be informed, so that a new shipment might be prepared and this opportunity not be lost, to avoid the memorable things of this New World from falling into oblivion.
> Bernardino de Sahagún, Letter to King Philip II, Mexico, 26 March 1578.

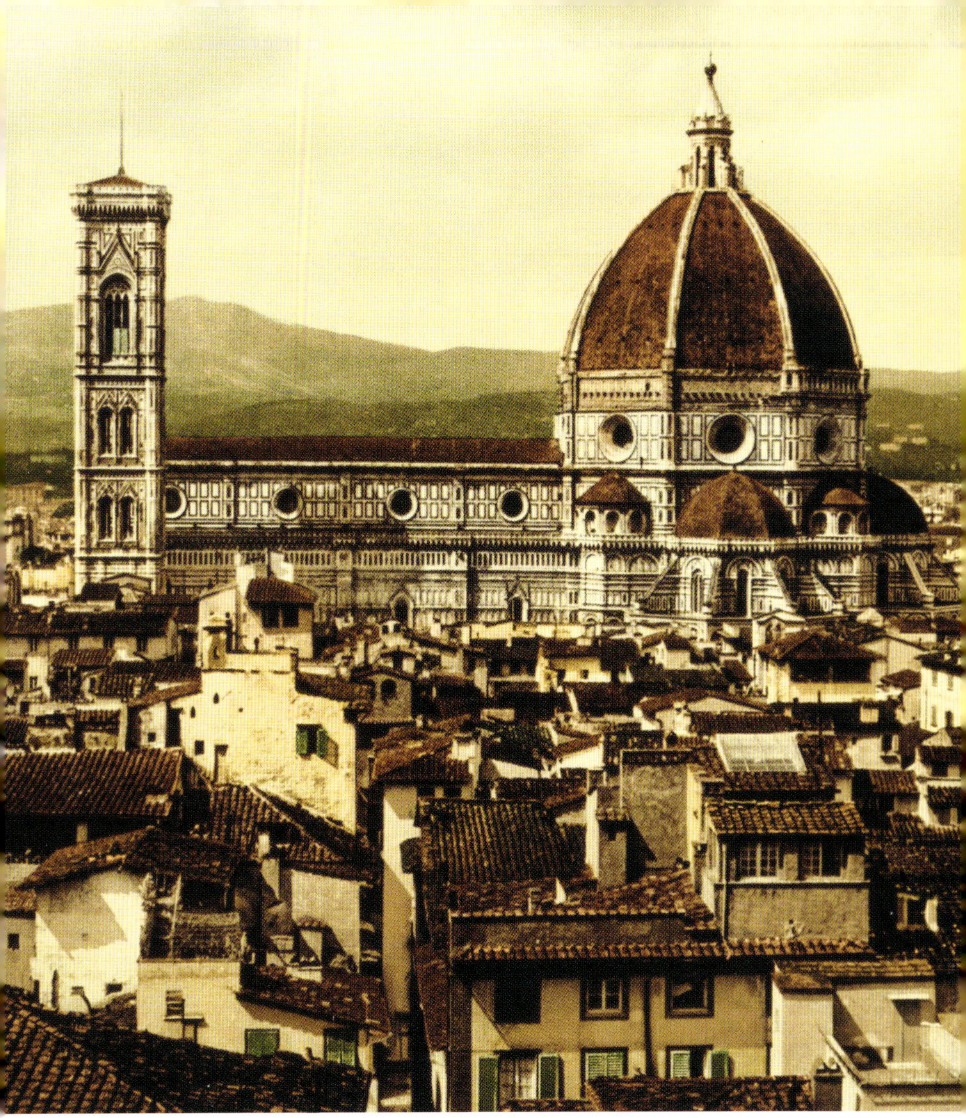

The cathedral of Santa Maria del Fiore in Florence as seen from Palazzo Vecchio.

The pre-Columbian world, although physically defeated, ultimately survived. History is here to prove that Florence is a leader in developing a culture around the preservation of the 'memorable things of this New World'. [R.V.]

In Florence at the end of the 16th century, the Medici dynasty distinguished itself for its particular attention to and curiosity about everything that came from the New World, as demonstrated by their extraordinary collections containing rare and precious pre-Columbian artefacts and objects. It is mainly thanks to the Medici that chocolate became popular in Italy. In 1670, Cosimo III de' Medici (1642–1723) succeeded his father as Grand Duke of Tuscany. This was at the height of the Baroque era, at a time when chocolate was a vice and a fashion throughout Europe, as reported by Learco Nencetti and Elisabetta Francioni on the occasion of the exhibition 'Cioccolata, squisita gentilezza' (2005) at the Florence Biblioteca Nazionale Centrale.

> … at the house of the Medici [chocolate] was sipped in the famous forty-six chocolate *chicchere*, personal possessions of Cosimo III, and in a special jasmine-scented recipe dictated by the Grand Duke himself (a drink of 'more exquisite delicacy' than that of the King of Spain) and which remained a 'state secret' for a long time.
> Elisabetta Francioni, 'La Cioccolata nei libri', *Biblioteche oggi. Rivista bimestrale di informazione ricerca e dibattito*, 23/2 (2005), pp. 92–3: 92.

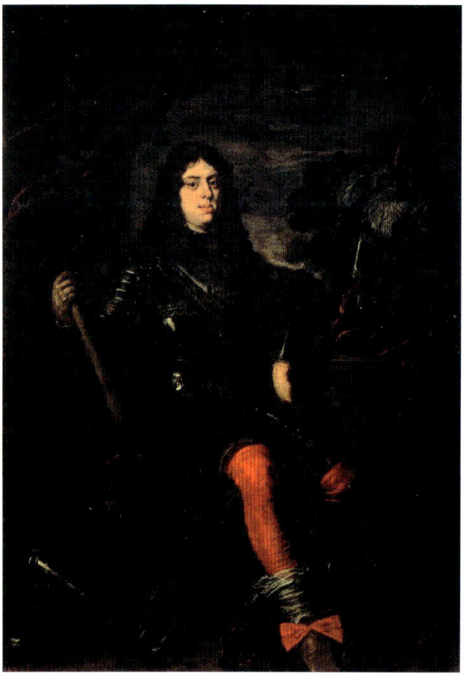

Giusto Sustermans, *Portrait of Cosimo III de' Medici*, 17th century, oil on canvas, Poggio a Caiano, Museum of the Medici Villa.

Grand Duke Cosimo III de' Medici was himself a passionate chocolatier. The recipe was developed by the archiater and superintendent of the Grand Ducal Foundry and Apothecary, the scientist and philologist Francesco Redi. It was secret, so secret that the Grand Duke made it a state secret, but it was handed down in a postscript to a letter by Redi, which was kept by Antonio Vallisneri (1661–1730) and published after the scientist's death (1697):

Take roasted cocoa, and coarsely crush 10 lbs. Take enough fresh jasmines to be mixed with the said cocoa, placing them layer upon layer in a box or other vessel, and let them stand for 24 hours, and then take them out and put as many again in the same cocoa, placing layer upon layer as before; and in this way, every 24 hours fresh jasmines are placed ten or twelve times. Then take 8 lbs of good dry white sugar. 3 oz of perfect vanilla. 6 oz of perfect fine cinnamon. 2 scruples of grey amber and in accordance with the art, chocolate is made; being careful when making it, that the stone be not too hot; but that the craftsman work it, that he does not pass four or five lbs per mass at the most, because if he were to heat the stone too much, it would lose its scent.
Francesco Redi, *Opere. II*, Venice, 1712, pp. 47–8.

Flower buds picked before dawn release their fragrance as they blend smoothly with the all-enveloping aroma of cocoa. Unlike the other recipes popular in Europe, that of the Medici was a flower-scented chocolate, just like the chocolate of the Mexica which was precisely transcribed in Sahagún's Book X. [R.V.]

> … significant collections of American artefacts were created in Italy: the richest being those of the Medici family in Florence.
> Francesca Cantù, *La Conquista spirituale. Studi sull'evangelizzazione del Nuovo Mondo*, Rome, 2007, p. 38.

When the House of Lorraine took over the government of Tuscany from the Medici in the 18th century, the director of the Biblioteca Laurenziana, Angelo Maria Bandini (1726–1803), while compiling the *Catalogus codicum latinorum* following the addition of codices ordered by Grand Duke Peter Leopold (1747–92), identified in 1793 a work described as an integral part of the Medici family's private collection of manuscripts, among the codices of the Biblioteca Medicea Palatina transferred from Palazzo Pitti, as follows: 'Historia mexicana libri dodici in lingua hispanica et mexicana exaratis'.

For reasons that are still historically mysterious, Sahagún's original work did not end up on the bonfire of the Inquisition, contrary to the infamous order of the Cédula Real issued by King Philip II of Spain. One of the most reliable hypotheses as to how Sahagún's immense anthropological work arrived in Florence is that it was the friar Rodrigo de Sequera himself, commissary general of the Franciscan Order and a great admirer of Sahagún's work, who smuggled the books into Europe and ensured their survival. At the beginning of 1580, Sequera left Mexico taking the manuscripts with him and probably after a stop in Spain went to Rome to see Cardinal Ferdinando de' Medici, the future Grand Duke of Tuscany, to deliver the precious volumes to him, thus saving them from the fate ordered by King Philip II.

The collecting of American artefacts developed in Italy in Florence from 1569 – the year of the establishment of the Grand Duchy of Tuscany by Cosimo I de' Medici (1519–74) – to around 1600, in the period corresponding to the rule of Francesco I (1541–87), Cosimo's son, and that of Ferdinand I. The Medici collections were created through purchases and gifts of rare and precious objects and artefacts from the New World. Although Cosimo and Francesco were fascinated by news of the new continent, Ferdinando was the only one to finance a voyage from Livorno to America, with the unique initiative of founding a Tuscan colony in the New World. During his years as a cardinal, he became a very influential figure, even holding the position of Protector of the Faith in Spain, of the Franciscan Order of Friars Minor, a title given to him by Pope Gregory XIII (1501–85). Ferdinando returned from Rome to Florence to assume the title of Grand Duke of Tuscany in 1587. It was then that he probably brought the precious handwritten volumes of Bernardino de Sahagún with him to Florence.

> In 1589 the Florentine painter Ludovico Buti (active 1560–1611) painted on a ceiling in the Uffizi [the Armoury] a fresco clearly inspired by the illustrations in the manuscript, a clear

sign that it had already become part of the Biblioteca Medicea Palatina, where it complemented the wealth of Mexican objects in the Medici Grand Dukes' collections of non-European origin …
Ida Giovanna Rao, 'Il Codice Fiorentino', in *Il mondo degli Aztechi nel Codice Fiorentino*, Florence, 2007, pp. 7–13: 8.

Friar Rodrigo's motives cannot be ascertained, but he left them with someone he knew would take care of them and appreciate them.
Ida Giovanna Rao, in Alejandra Ortiz Castañares, 'El Códice Florentino ya puede consultarse con un clic', *La Jornada*, 6 February 2013.

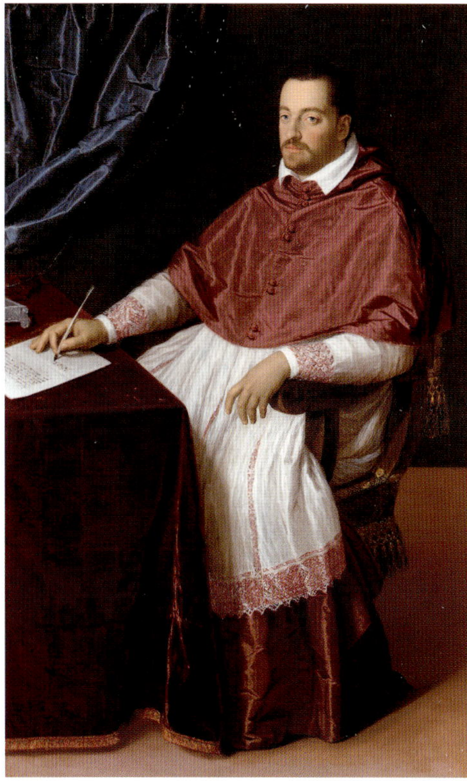

Scipione Pulzone, *Ritratto del cardinale Ferdinando de' Medici*, 1580, oil on canvas, 185 × 119.3 cm, Adelaide, Art Gallery of South Australia.

The twelve books divided into three volumes corresponding to manuscripts nos. 218–20 in the Medicea Palatina collection of the Biblioteca Medicea Laurenziana in Florence are the final and original transcription of the *Historia general de las cosas de Nueva España* by Bernardino de Sahagún, known universally for this reason as the Florentine Codex.

It was probably the Florentine Codex that inspired Grand Duke Cosimo III and the erudite Redi to use precious flowers (unheard of in European courts) in the preparation of chocolate, as was customary in the court of the Mexican princes and nobles:

Aztec royalty and nobility were devoted to music, song, dance, and poetry. … In the latter [poems], chocolate – especially xochicacahuatl, 'flowered chocolate' – was a metaphor for luxuriousness and sensuality.
Sophie D. Coe, Michael D. Coe, *The True History of Chocolate*, London, 1996, p. 103. [R.V.]

LEGADO

> With the Cédula Real of 22 April 1577, Philip II prohibited any compendium of the monumental work of Bernardino de Sahagún, and any other writing concerning the memory of ancient Mexico. A cultural genocide by decree … And the haunting ambiguity one feels when reading them, leaves one wondering whether this is a conscious chapter in the battle of memory against the darkness of oblivion.
> Vittoria Martinetto, 'Introduction' to Bernardino de Sahagún, *I colloqui dei Dodici*, Palermo, 1991, p. 9.

The journeys of the Sahagún manuscripts represent one of the most intriguing episodes in Mexican bibliography. Predating the *Historia general de las cosas de Nueva España*, there are several books, outlines and drafts that, taken together, constitute the surviving and preserved texts of Bernardino de Sahagún's work. In particular, four main manuscripts stand out: the Florentine Codex, the two Códices Matritenses and the Toulouse Codex.

The twelve books containing texts in Nahuatl and Spanish arranged in two columns and accompanied by more than 2000 illustrations, compiled from 1575 to 1577, known as the Florentine Codex and preserved in the Biblioteca Medicea Laurenziana in Florence, are the final and original version of the *Historia general de las cosas de Nueva España*, delivered to Sequera following the Cédula Real of King Philip II. For this reason, it is also sometimes referred to as the 'Sequera Manuscript'.

Instead, the manuscripts handed over to Viceroy Martín Enriquez, later sent to Phillip II, are those corresponding to the drafting of the twelve books in the Nahuatl language completed in 1565, and redrafted in 1569, from which the Florentine Codex was born. They are known today as the Códices Matritenses because they are stored in Madrid. They include the *Primeros Memoriales* and the *Manuscrito de Tlatelolco*, consisting of *Primer manuscrito de Tlatelolco*, also known as *Segundos Memoriales*), *Memoriales en tres columnas* and *Memoriales con escolios*. Preserved at the Biblioteca del Palacio Real (books I to VI) and at the Real Academia de Historia (books VIII to XI, as well as some sheets of the draft of the 'third column') contain practically the entire work between the two, with the exception of books VII and XII. Francisco del Paso y Troncoso (1842–1916), Mexican philologist, linguist, historian and archaeologist, reported that the Códices Matritenses in themselves consist of three volumes, two of them written in Mexican, with illustrations, and the third in Castilian, without illustration. This Spanish version, found in the third volume, is supposed to be a copy extracted from Sequera's copy in Spain. It is worth noting that Sahagún had conceived the idea of including a 'third column' dedicated to explaining the terms in Nahuatl that he considered relevant. Unfortunately, he was unable to complete this project.

The Toulouse Codex remained in safekeeping in the convent of San Francisco de Tolosa in Navarre until 1783. A copy of the Spanish texts of the Florentine Codex, without illustrations, can be found today at the Real Academia de la Historia. Almost all the Spanish editions, whether complete or in fragments, derive indirectly from the Toulouse manuscript through one of its copies: the 1793 copy, made by Diego García Panes y Abellán (1734–1801); the 1802 copy, destined for the Franciscans of Toulouse, in exchange for the original; the 1815 copy, belonging to Felipe Bauzá (1764–1834) and published in 1830 by Lord Kingsborough (Edward King, 1795–1837). Panes y Abellán brought his copy to Mexico in 1795, and several Mexican editions were derived from it: in 1829/30 by Carlos María de Bustamante (1774–1813), in 1886 by Joaquín García Icazbalceta (1824–94), in 1938 by Joaquín Ramírez Cabañas (1886–1945), in 1946 by Miguel Acosta Saignes (1908–89) and in 1956 by Ángel María Garibay Kintana (1892–1967).

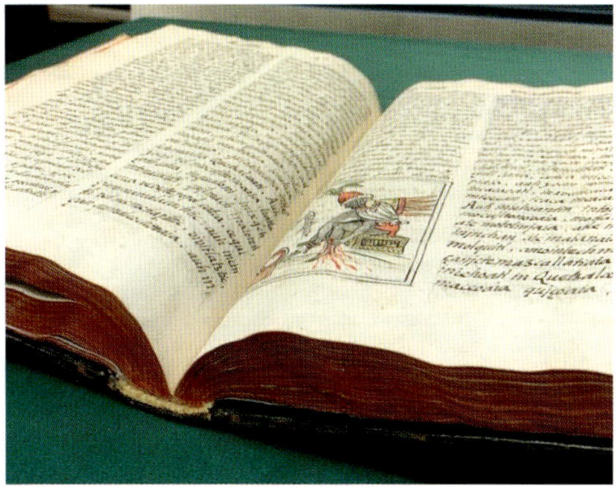

The Florentine Codex, display case no. 19 of the exhibition at the Biblioteca Medicea Laurenziana.

The first complete transcription of the Florentine Codex was made by del Paso y Troncoso in the 1890s. Later, in 1905, he promoted the chromolithographic reproduction published by Hauser y Menet in Madrid. In an interview given in 2013, Ida Giovanna Rao recalls that, 'The real discovery of the codex began with the manuscript of the Mexican scholar Francisco del Paso y Troncoso, who came here (Biblioteca Medicea Laurenziana, [A/N.]) and transcribed everything, book by book. Later, his texts were lost: of the 12 transcribed books, six are kept at the National Museum of Anthropology in Mexico City, while the others are missing'.

In 1961, an edition that includes a complete palaeography of the column in Nahuatl was published with the title *Florentine Codex. A General History of the Things of New Spain. Translated from the Aztec into English*, by Arthur J. O. Anderson and Charles E. Dibble, published by the University of Utah, in Santa Fe.

In 1979, the first facsimile edition of the Florentine Codex was published on the initiative of the Mexican government through the Secretería de Gobernación and the Archivo General de la Nación, published by the Giunti-Barbera Publishing House in Florence.

In 1988, Alfredo López Austin and Josefina García Quintana produced a renewed edition of the full Spanish-language text of the Florentine Codex, which was published in three volumes by Alianza Editorial. Two years later, Conaculta (the Mexican National Council for Culture and the Arts) published a version that included a glossary of Nahuas terms and archaic or uncommon Spanish words. In 2001, the publishing houses Más Cultura and Aldus published a luxury, four-volume facsimile edition of the Florentine Codex, edited by Miguel León-Portilla.

Since November 2012, the Florentine Codex has been officially included in the collections of the World Digital Library, thanks to the collaboration between the United States Library of Congress and the Biblioteca Medicea Laurenziana. The Florentine Codex, as Vera Valitutto pointed out in 2013, represents a heritage 'available not only to scholars but to anyone interested in learning more about this incredible civilisation and this important chapter in human history'.

On 26 October 2023, the Getty Research Institute presented the Digital Florentine Codex project at the Getty Centre in Los Angeles, and on 2 November 2023 at the Biblioteca Medicea Laurenziana in Florence, at the opening of the exhibition 'Cacao Entre Dos Mundos'. The project, led by the Getty Research Institute and UCLA's Latin American Institute, offers a digitised version that facilitates thematic or book-based exploration. It also permits the user to search for words and images among the 2500 pages of the precious manuscript. The texts are transcribed in the original Nahuatl and Castilian, and accompanied by translations into English, Spanish and the Nahuatl of the Huasteca.

> The indigenous legacy in Mexico is alive and thriving; failure to acknowledge and appreciate it means denying and ignoring one's present.
> Kim Richter, in Alejandra Ortiz Castañares, 'Está lista una versión digitalizada más accesible del Códice Florentino', *La Jornada*, 18 November 2023. [R.V.]

> In Tlatelolco, during the 16th century, the intellectual 'mestizaje' of two cultures was born: the indigenous and the European. Two worlds that could no longer deny each other's existence.
> Baltazar Brito Guadarrama, in Martín De La Cruz, Juan Badiano, *Libellus de Medicinalibus Indorum Herbis* [1552], facsimile reprint, Mexico, 2020, p. 19.

Before Sahagún, no one had devoted themselves to exploring in detail a culture that was considered new and radically different. Friar Bernardino believed that what was heard and then documented, even if it came from the most modest man, could have universal meaning and value. There is another great innovation in his work: he concludes with a unique text, the account of a contemporary event, the conquest of Mexico, narrated by the voices of the defeated.

> Sahagún, by directing his vast encyclopaedia towards the comprehension of a language and culture that was new and radically different from his own, overcame the ethnocentrism of the Western tradition by creating an innovative way of understanding the human being in its actions, which we now call anthropology. Anthropology helps us all understand that there are no inferior languages or cultures and that every human being and each population is a protagonist in the universe of cultures.
> Ascensión Hernandez de León Portilla, 'La Historia General de Sahagún a la luz de las enciclopedias de la tradición greco-romana', in *Bernardino de Sahagún: quinientos años de presencia*, edited by Miguel León-Portilla, México, 2002, pp. 41–59: 59.

In 1992, UNESCO established the 'Memory of the World' programme with the aim of contributing to the preservation of humanity's rich documentary heritage and preventing the irrevocable loss of collective memory, in order to facilitate its access and dissemination and raise public awareness of its significance and the need to preserve it. Since 2015, the Bernardino de Sahagún Florentine Codex has been included in the International Register of the Memory of the World.

> The preservation and dissemination of documentary heritage is vital for greater understanding and dialogue between peoples in order to promote peace, human rights and dignity.
> 'Memory of the World' programme, 1992.
> [R.V.]

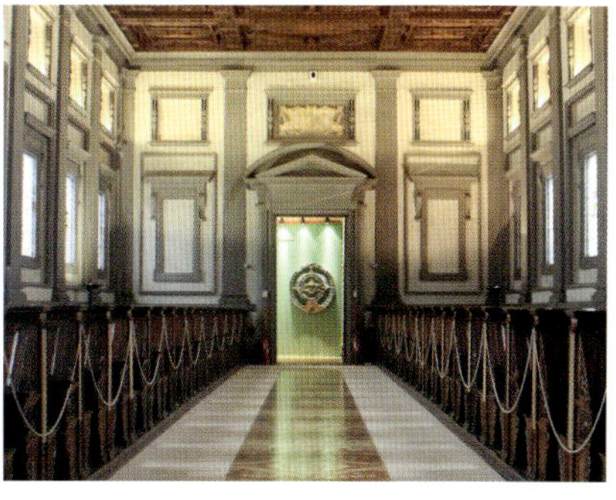

The entrance to the exhibition at the Laurenziana.

Antonio Rodriguez, *Portrait of Moctezuma II*

second half of 17th century
oil on canvas, 182 × 106.5 cm
Florence, Uffizi Galleries, Palazzo Pitti, Tesoro dei Granduchi, inv. 1890, no. 5158

> The basis for genuine sustainable development emerges from this play of contemporaneity, in which past and present are reconciled.
> Alejandra Moreno Toscano, in Martín De La Cruz, Juan Badiano, *Libellus de Medicinalibus Indorum Herbis* [1552], facsimile reprint, México, 2020, p. 8.

Emperor Moctezuma II is depicted in all his magnificence in this painting that has been associated with a portrait in a Codex. It is likely that the Grand Duke of Tuscany, Cosimo III de' Medici, received it in 1690 or shortly before, presumably through the intermediary of the Franciscan Order. Quite fortuitously, the protagonists of a narrative full of historical convergences, where history is still being written, cross paths in this painting.

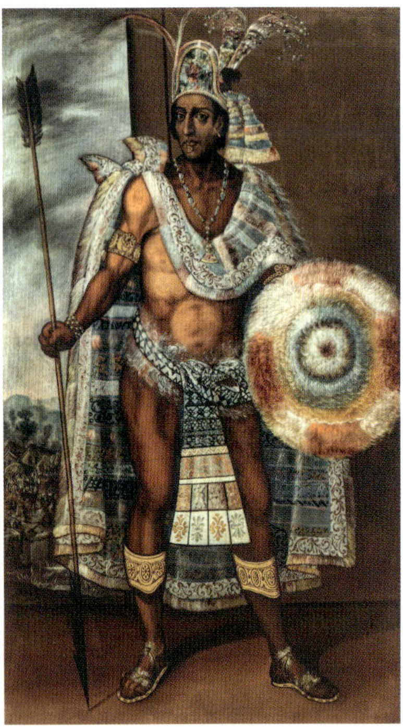

For five hundred years, the Florentine Codex and cocoa have travelled a path of sweet fatality and unjust bitterness. Unfortunately for cocoa, its history still continues on a troubled path, which hopefully, however, will take a positive turn. Just like the unexpected happy epilogue of the troubled affairs of Moctezuma II and Cortés.

> After the death of Moctezuma and the conquest of Mexico, Emperor Charles V gave Hernán Cortés the marquisate of the Oaxaca Valley and for a time he lived as a great lord. He did not, however, end his days in glory. Returning to Spain and plagued by conspiracies, he died in 1547 near Seville, but at his own request his remains were taken back to Mexico. He wanted to be buried in Huitzilan, the site where he first met Montezuma on a November day in 1519.

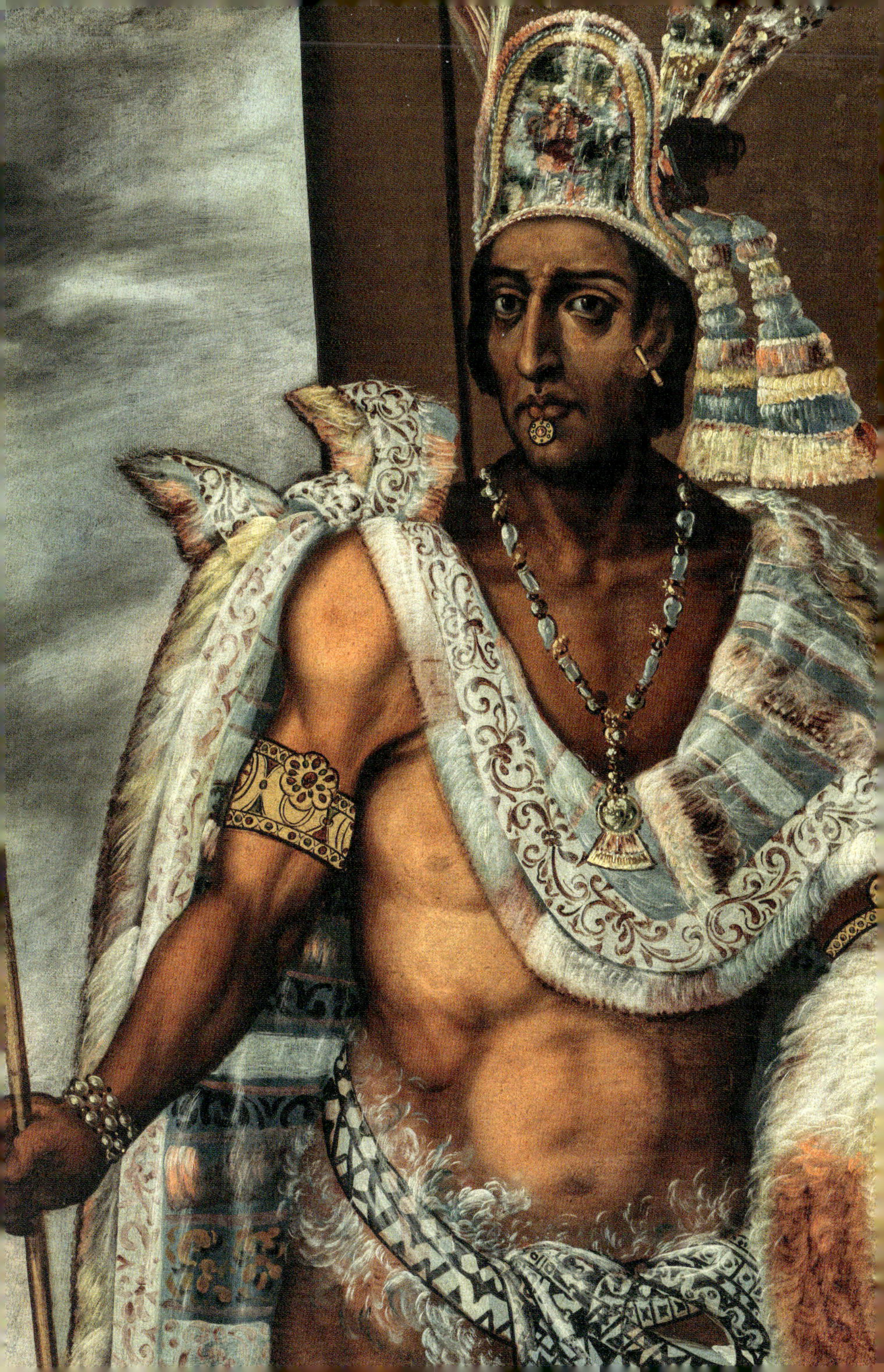

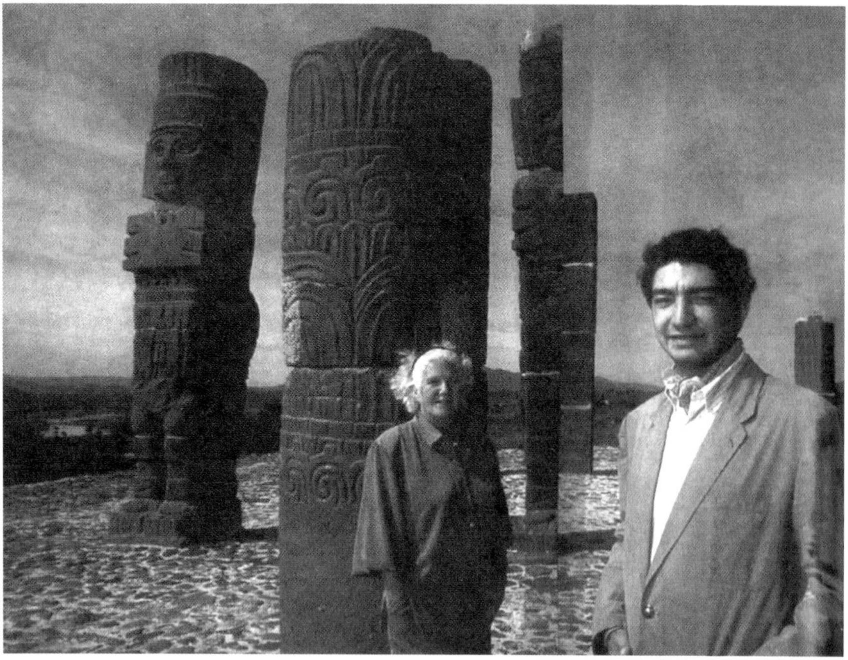

The meeting between Maria Gloria Aragona Pignatelli Cortés and Juan José Manuel Marcilla de Teruel Moctezuma, Tula, 12 October 1992.

> Centuries later, 500 years after the discovery of the New World, a unique event happened on 12 October 1992: Maria Gloria Aragona Pignatelli Cortés, the last direct descendant of Hernán Cortés, and Juan José Manuel Marcilla de Teruel Moctezuma, the last descendant of Montezuma, met in Tula in a gesture of historic reconciliation, the very place where according to legend Quetzalcóatl came down from heaven to give the cacao tree to mankind. Ruby Villarreal, *Squisita scoperta, il granducato di Toscana e la via del Cioccolato*, fact sheet, December 2012.

An auspicious sign of a happy ending for the history of chocolate. [R.V.]

Bibliography

Publications

Alma Flor Ada, *Vuelo del Quetzal*, Miami, FL-Santillana, 2002.

Antonio Aimi, *La "verdadera" visión de los vencidos. La conquista de México en las fuentes aztecas*, Alicante, 2009.

Celeste Alvarez Lopez, *Piedra del Sol*, Mexico, 2021.

Jorge Amado, *Cacau*, Rio de Janeiro, 1933; new ed. Caracas, 1991.

Maria Gloria Aragona Pignatelli Cortés, *Twilight at the Hacienda Cortés*, Brighton, 2010.

Gilberto Aranda Bustamante, Sergio Salinas Cañas, *Conflictos de identidades y politica internacional*, Santiago, 2005.

Luis Arranz Márquez, *Cristóbal Colón: Misterio y Grandeza*, Madrid, 2006.

Amalia Attolini Lecón, 'Cuentas, dares y tomares del cacao: delicia, convite, rito mesoamericano. Aspectos antropológicos', *Revista Digital Universitaria UNAM*, 12/4 (2011), 19 pp.

Amerigo Vespucci e i mercanti viaggiatori fiorentini del Cinquecento, edited by Margherita Azzari and Leonardo Rombai, Florence, 2013.

Oliver Balch, 'Mars, Nestlé and Hershey to face child slavery lawsuit in US', *The Guardian*, 12 February 2021.

Manuel Barbero Richart, 'Los Códices Etnográficos: el "Códice Florentino"', *Estudios de Historia Social y Económica de América*, no. 14 (1997), pp. 349–79.

Alfredo Barrera Vázquez, 'La Cultura Maya', *Anales del Museo Nacional de Arqueología, Historia y Etnografía*, 5th series, 2 (1935), pp. 327–49.

Juan José Batalla Rosado, 'Prisión y muerte de Motecuhzoma, según el relato de los códices mesoamericanos', *Revista Española de Antropología Americana*, 26 (1996), pp. 101–20.

Georges Baudot, 'Felipe II Frente a Las Culturas y a Los Discursos Prehispánicos de América. De La Transculturación a La Erradicación', *Caravelle. Cahiers du monde hispanique et luso-brésilien*, no. 78 (2002), pp. 37–56.

Bonnie K. Bealer, Bennet A.Weinberg, *The World of Caffeine: The Science and Culture of the World's Most Popular Drug*, New York 2001.

Giuseppe Bellini, *Storia della civiltà spagnola e americana*, Milan, 1957.

Giuseppe Bellini, *Tra Medioevo e Rinascimento. La poesia nell'America conquistata*, Salerno, 2003.

Paolo Broggio, 'Le congregazioni romane e la confessione dei neofiti del Nuovo Mondo tra "facultates" e "dubia": riflessioni e spunti di indagine', *Mélanges de l'École Française de Rome. Italie et Méditerranée*, 121/1 (2009), pp. 173–97.

Paolo Cabras, Aldo Martelli, *Chimica degli alimenti: nutrienti, alimenti di origine vegetale, alimenti di origine animale, integratori alimentari, bevande, sostanze indesiderabili*, Padua, 2004.

Serena Cadamuro, *Globalizzazione e diritti umani: lo sfruttamento di lavoro minorile nelle piantagioni di cacao in Costa d'Avorio e Ghana*, MA thesis, Venice, Università Ca' Foscari, 2014.

Filippo Camerota, *Una terra oltre le stelle. Il planisfero di Martin Waldseemüller (1507)*, Florence, 2016.

Francesca Cantú, *La Conquista spirituale. Studi sull'evangelizzazione del Nuovo Mondo*, Rome, 2007.

Roberto Caraceni, *La degustazione del cioccolato: degustazione, valutazione, analisi organolettica*, Milan, 2010.

Francesco Carletti, *Ragionamenti … sopra le cose da lui vedute ne' suoi viaggi sì dell'Indie Occidentali, e Orientali come d'altri paesi*, ed. Florence, 1701.

Francesco Carniel, *La piuma e l'acciaio. Le cause di una sconfitta impossibile*, MA thesis, Venice, Università Ca' Foscari, 2014.

Laura Carrillo Marquez, 'Tesoros de piedra: Estudio de procedencia y tecnología del Arte Indígena en las Cámaras de Maravillas de los Medici de Toscana en los Siglos XVI, XVII y XVIII', *CLÍO. Arqueológica*, 35/2 (2020), pp. 1–28.

Bartolomé de Las Casas, *Los indios de México y Nueva España*, ed. Mexico, 1966.

Noam Chomsky, *Year 501: The Conquest Continues*, London, 1993.

Fernando Ciaramitaro, 'Gli italiani del Nuovo Mondo: conquistatori, colonizzatori ed evangelizzatori', in *El mundo de los conquistadores*, edited by Martín F. Ríos Saloma, México, 2015, pp. 631–62.

Cioccolata, squisita gentilezza, catalogue of the exhibition (Florence, Biblioteca Nazionale Centrale, 2005), edited by Piero Scapecchi and Learco Nencetti, Florence, 2005.

Il cioccolato dai Maya al XX secolo, catalogue of the exhibition (Alba, Fondazione Ferrero, 2008–2009), edited by Andreina D'Agliano, Cinisello Balsamo, 2008.

Sophie D. Coe, Michael D. Coe, 'Estampas del Chocolate en Europa', *Artes de México*, no. 105 (2011), pp. 53–62.

Sophie D. Coe, Michael D. Coe, *The True History of Chocolate*, London, 1996; Spanish translation, Mexico, 2013.

Carlo Collodi, *Le avventure di Pinocchio. Storia di un burattino*, Florence, 1883.

El conquistador anónimo: Relación de algunas cosas de la Nueva España y de la gran ciudad de Temestitán México escrita por un compañero de Hernán Cortés, edited by Leon Díaz Cárdenas, Mexico, 1941.

Hernán Cortés, *Cartas de relación de la conquista de México* [1519, 1520, 1522, 1524, 1526], ed. Madrid, 1970.

David Cuevas Gongora, 'El tesoro perdido de Moctezuma', *Baetica. Estudios de Arte, Geografía e Historia*, 33 (2011), pp. 283–98.

Ann Cyphers, 'La detección de teobromina en vasijas de cerámica olmeca', *Arqueología*, 46 (1987), pp. 153–66.

Jaqueline De Durand-Forest, 'El Cacao entre los Aztecas', *Estudios de cultura Náhuatl*, no. 7 (1967), pp. 115–81.

Martín De La Cruz, Juan Badiano, *Libellus de Medicinalibus Indorum Herbis* [1552], facsimile reprint, Mexico, 2020.

María Del Carmen León Cázares, 'Fray Diego de Landa, obispo electo de Yucatán, según su información de legitimidad y nobleza de 1572', *Estudios de Cultura Maya*, no. 58 (2021), pp. 295–328.

Gloria M. Delgado De Cantú, *Historia de México. I*, Mexico City, 1993.

Bernal Díaz Del Castillo, *Historia verdadera de la conquista de la Nueva España* [1632], edited by Miguel León-Portilla, Madrid, 1984; new ed. Mexico, 1993.

Carlos D'Ors, 'Quetzalcóatl y Coatlicue', *Cuadernos Hispanoamericanos*, no. 449 (1987), pp. 7–28.

Dolceamaro. Storia e storie dal cacao al cioccolato, edited by Zeffiro Ciuffoletti, Florence, 1997.

Diego Durán, *Ritos y fiestas de los antiguos mexicanos*, Mexico City, 1980.

Enrique D. Dussel, *The Invention of the Americas. Eclipse of "the Other" and the Myth of Modernity*, New York, 1995.

Umberto Eco, 'La leggenda della terra piatta', *La Repubblica*, 23 February 2009.

Philippe Erlanger, *Charles Quint*, Paris, 1980.

Laura Esquivel, *Como agua para chocolate*, Mexico City, 1989.

Gonzalo Fernández de Oviedo y Valdés, *Historia natural y general de las Indias, islas y tierra firme del mar océano* [16th century], 1st ed., 4 vols., Madrid, 1851–5.

Laura Filloy Nadal, 'El Jade en Mesoamerica', *Arqueología Mexicana*, no. 133 (2015), pp. 30–6.

Laura Filloy Nadal, 'Una mirada al pasado desde el presente', *Arqueología Mexicana. Ediciones Especiales*, no. 111 (2022), pp. 10–11.

Jean-Louis Flandrin, Massimo Montanari, *Storia dell'alimentazione*, Rome–Bari, 1997.

Enrique Florescano, 'La Saga De Ce Ácatl Topiltzin Quetzalcóatl', *Relaciones. Estudios de historia y sociedad*, no. 95, 24 (2003), pp. 201–34.

Elisabetta Francioni, 'La Cioccolata nei libri', *Biblioteche oggi. Rivista bimestrale di informazione ricerca e dibattito*, 23/2 (2005), pp. 92–3.

Bernardo García Martinez, 'El Marquesado del Valle', *Arqueología Mexicana*, no. 153 (2018), pp. 68–73.

José García Payón, *Amaxocoatl o Libro de Chocolate*, Toluca, 1936.

Antonio García y García, 'Los privilegios de los religiosos en la evangelización de América', *Mar oceana. Revista del humanismo español e iberoamericano*, nos. 11–12 (2002), pp. 45–64.

Ángel M. Garibay Kintana, *Historia de la literatura náhuatl*, México, 1954.

Geschichte der Welt Die Welt vor 600: Frühe Zivilisationen, edited by Hans-Joachim Gehrke, Munich 2017.

Rosalba Gioffré, *Cioccolato: nuove armonie: cioccolatini, torte e crostate, piatti salati, cocktail, dolci al cucchiaio*, Florence, 2003.

Alessandro Giraudo, *Storie straordinarie delle materie prime*, Turin, 2019.

Martín González de la Vara, 'Cacao. 1500 años en Tabasco', *Arqueología Mexicana*, no. 12 (1995), pp. 63–6.

Michel Graulich, 'Los reyes de Tollan', *Revista Española de Antropología Americana*, 32 (2002), pp. 87–114.

Detlef Heikamp, *Il Nuovo Mondo*, in *Magnificenza alla Corte dei Medici. Arte a Firenze alla fine del Cinquecento*, catalogue of the exhibition (Florence, Palazzo Pitti, Museo degli Argenti, 1997–8), Milan, 1997, pp. 399–402.

Francisco Hernandez, *Obras completas*, 7 vols., 1960–84, Mexico, 2015.

Ascensión Hernandez de León Portilla, 'La Historia General de Sahagún a la luz de las enciclopedias de la tradición greco-romana', in *Bernardino de Sahagún: quinientos años de presencia*, edited by Miguel León-Portilla, Mexico, 2002, pp. 41–59.

Kenneth Hirth, 'Xochicalco durante el Epiclásico', *Arqueología Mexicana*, no. 153 (2018), pp. 47–53.

Piero Innocenti, *La geografia oggi. Individui, società, spazio*, Rome, 2001.

David L. Katz, Kimberly Doughty, Ali Ather, 'Cocoa and Chocolate in Human Health and Disease', *Antioxidants and Redox Signaling*, 15 (2011), pp. 2779–811.

Introducción a los Jeroglíficos Mayas, XVI Conferencia Maya Europea Copenhague, edited by Harri Kettunen and Christophe Helmke, Copenhagen 2011.

Katherine Khodorowsky, Olivier De Loisy, *Chocolat et grands crus de cacao*, Paris, 2003.

Joseph Ki-Zerbo, *Histoire de l'Afrique noire*, Paris, 1972.

Herbert S. Klein, *The Atlantic Slave Trade*, Cambridge, 2010 (2nd ed.; 1st ed. Cambridge, 1999).

Jacques Lafaye, *Quetzalcóatl et Guadalupe. La formation de la conscience nationale au Mexique*, Paris, 1974.

Luis Leal, 'El libro XII de Sahagún', *Historia Mexicana. El Colegio de México*, no. 18, 5/2 (1955), pp. 184–210.

Miguel León-Portilla, *Los antiguos mexicanos a través de sus crónicas y cantares*, Mexico, 1961.

Miguel León-Portilla, 'Atlaquetzalli: agua preciosa', *Artes de México*, no. 105 (2011), pp. 34–9.

Miguel León-Portilla, *Bernardino de Sahagún. Pionero de la antropología*, Mexico, 1999.

Miguel León-Portilla, 'La historia bilingüe de Sahagún: ¿existió un Manuscrito Enríquez además del Códice Florentino?', *Estudios de cultura Náhuatl*, no. 21 (1991), pp. 133–44.

Miguel León-Portilla, 'El retorno de Quetzalcóatl', *Arqueología Mexicana*, no. 53 (2002), pp. 54–7.

Massimo Livi Bacci, *Conquista. La distruzione degli indios americani*, Bologna, 2005.

Maria Longhena, *Antico Messico. Storia e cultura dei Maya, degli Aztechi e di altri popoli precolombiani*, Vercelli, 1998.

Alfredo López Austin, 'La cosmovisión de la tradición mesoamericana. Tercera parte', *Arqueología Mexicana*, special ed., no. 70 (2016).

Alfredo López Austin, Leonardo López Luján, *Mito y realidad de Zuyuá. Serpiente Empluma-*

da Y Las Transformaciones Mesoamericanas del Clásico al Posclásico, Mexico City, 1999.

Alfredo López Austin, Leonardo López Lujan, 'Tollan y su gobernante Quetzalcóatl', *Arqueología Mexicana*, no. 67 (2004), pp. 38–43.

Leonardo López Luján, Olivier Guilhem, 'Images of Moctezuma and His Symbols of Power', in *Moctezuma: Aztec ruler*, catalogue of the exhibition (Londra, British Museum, 2009–10), edited by Colin Mcewan and Leonardo López Luján, London, 2010, pp. 2–13.

Diana Magaloni Kerpel, 'Painters of the New World: The Process of Making the "Florentine Codex"', in *Colors between Two Worlds. The Florentine Codex of Bernardino de Sahagún*, proceedings of the symposium (Florence, 12–13 June 2008), Milan, 2011, pp. 47–76.

Pier Giacomo Magri, 'L'espansione degli stati iberici in Africa ed America e la questione del patronato reale', *Studi Urbinati, A - Scienze Giuridiche, Politiche ed Economiche*, 52/2 (2001), pp. 209–54.

Charles C. Mann, *1493: Uncovering the New World Columbus Created*, New York, 2011.

Los manuscritos de la Historia general de las cosas de la Nueva España de Bernardino de Sahagún, el Códice Matritense de la Real Academia de la Historia, Madrid, 2013.

Giovanni Marchetti, 'Hacia la edición crítica de la Historia de Sahagún', *Cuadernos Hispanoamericanos*, no. 396 (1983), pp. 505–40.

Simon Martin, 'Cacao in Ancient Maya Religion', in *Chocolate in Mesoamerica: A Cultural History of Cacao*, edited by Cameron L. McNeil, Gainesville, 2008, pp. 154–83.

Eduardo Matos Moctezuma, *Los aztecas*, Milan, 1989.

Eduardo Matos Moctezuma, '¿Cuáles fueron las causas que incidieron en la caída de Tenochtitlan?', *Arqueología Mexicana*, no. 134 (2015), pp. 86–7.

Eduardo Matos Moctezuma, '¿Cuántos códices prehispánicos hay en México?', *Arqueología Mexicana*, no. 130 (2014), pp. 86–7.

Eduardo Matos Moctezuma, '¿Fue traidora la Malinche?', *Arqueología Mexicana*, no. 115 (2012), pp. 88–9.

Eduardo Matos Moctezuma, *Muerte a filo de Obsidiana*, Mexico City, 1975.

Eduardo Matos Moctezuma, 'Quetzalcóatl ¿blanco y de ojos azules?', *Arqueología Mexicana*, no. 113 (2012), pp. 82–3.

Eduardo Matos Moctezuma, '¿Venció Cortés con 800 españoles a miles de indígenas?', *Arqueología Mexicana*, no. 116 (2012), pp. 88–9.

Elsa Mazzolini, Alessandra Meldolesi, *L'Italia del cioccolato*, Milan, 2004.

Josè Luis Méndez, David Vilchis, 'Pioneros de los estudios superiores en el continente americano: fray Bernardino de Sahagún y el Colegio de la Santa Cruz de Tlatelolco', *Otros diálogos*, no. 16 (2021), 15 pp.

Enrique Méndez Bolaina, Mariana Diosdado Contreras, 'Cacao manjar de Dioses', *Ciencia y Luz*, 18 September 2018.

Mitologia del nuovo mondo. Miti e leggende dell'Oceania e delle Americhe, edited by Alice Mills, Milan, 2008.

Moctezuma II. Tiempo y destino de un gobernante, edited by Leonardo López Luján and Colin McEwan, Mexico City, 2010.

Rubén Bernardo Morante López, 'El ciclo de Venus en el retorno de Quetzalcóatl', *Cosmovisiones/Cosmovisões*, 3 (2021), pp. 195–228.

Fr. Toribio Motolinía de Benavante, *Memoriales e Historia de los indios de la Nueva España* [1536], ed. Madrid, 1970.

Van Tang Nguyen, Nghia Huu Nguyen, 'Proximate Composition, Extraction, and Purification of Theobromine from Cacao Pod Husk (Theobroma Cacao L.)', *Technologies*, 5/2 (2017), 14 pp.

Nisao Ogata, *Domestication and Distribution of the Chocolate Tree (T. cacao L.) in México*, in *Lowland Maya Area. Three Millenia at The Human-Wildland Interface*, edited by A. Gómez-Pompa et al., New York, 2003, pp. 415–33.

Guilhem Olivier, Leonardo Luján López, 'Las imágenes de Moctezuma II y sus símbolos de poder', in *Moctezuma II. Tiempo y destino de un gobernante*, catalogue of the exhibition (Mexico City, Museo de Templo Mayor, 2010), edited by Leonardo López Luján and Colin McEwan, Mexico, 2010, pp. 79–91.

Juan Francisco Ornelas, 'Origen y evolución de los colibríes', *Ciencias*, no. 42 (1996), pp. 38–47.

Raimondello Orsini, *Etica ed Economia: alcune riflessioni sul concetto di scelta*, Working Paper no. 33 (2006), Dipartimento di Scienze Economiche, Università di Bologna and AICCON.

Gigi Padovani, 'Il 'Cibo degli Dei' e i Savoia', *Civiltà della Tavola. Accademia Italiana della Cucina*, no. 349 (June 2022), pp. 4–5.

Margarita Palacios Sierra, 'Biografía y mito de Quetzalcóatl', *Anuario de Letras*, 19 (1981), pp. 163–81.

Miguel Pastrana Flores, 'Paleografía y traducción del náhuatl al español de "Del principio que tuvieron los dioses", párrafo primero del libro tercero del Códice florentino', *Estudios de Cultura Nahuatl*, no. 50 (2015), pp. 371–81.

José García Payón, *Amaxocoatl o libro de chocolate*, México, 1936.

Antonio Xavier Pérez, 'Teatro de la legislacion universal de España e Indias', in *Leyes de Recopilaciòn*, 28 vols., Madrid, 1791–8, 27 (1798), p. 421.

Antonio Pérez Henares, *La Española. Una isla en el Caribe fue el origen de todo un imperio*, Madrid, 2023.

Michael Pulvini, *The Dark Side of Chocolate. Storia delle frodi sul cioccolato*, MA thesis, Padua, Università degli Studi, 2012.

Giovanni Battista Ramusio, *Delle navigationi et viaggi*, 3 vols., Venice, Tommaso Giunti, 1550–9; ed. edited by Marica Milanesi, 6 vols., Turin, 1978–88.

Ida Giovanna Rao, *Mediceo Palatino 218-220 of the Biblioteca Medicea Laurenziana of Florence, Colors between Two Worlds. The Florentine Codex of Bernardino de Sahagún*, proceedings of the symposium (Florence, 12–13 June 2008), Milan, 2011, pp. 27–45.

Ida Giovanna Rao, Diana Magaloni, Alessandra Pecci, *Il mondo degli Aztechi nel Codice Fiorentino*, Florence, 2007.

Francesco Redi, *Opere. II*, Venice, 1712.

Victoria Ríos Castaño, 'From the "Memoriales con escolios" to the Florentine Codex: Sahagún and his Nahua assistants' co-authorship of the Spanish translation', *Journal of Iberian and Latin*, 20 (2014), pp. 214–18.

Patricio Robles Gil, Eugenia Pallares, 'El quetzal', *Arqueología Mexicana*, no. 15 (1995), pp. 54–9.

Kethlen Rodrigues, *I Medici e l'America: arte, etnografia e mitologia nell'immagine del nuovo mondo tra Cosimo I e Ferdinando I*, MA thesis, Verona, Università degli Studi, 2020.

Ángel Rosenblat, *La poblacion de America en 1492: viejos y nuevos calculos*, Mexico City, 1967.

Bernardino de Sahagún Ribera, *Historia general de las cosas de Nueva España*, MS, Tlatelolco, 1577, Mexico; edited by Ángel M. Garibay Kintana, Mexico, 1956; edited by Alfredo López Austin and Josefina García Quintana, 2 vols. Madrid, 1988.

Bernardino de Sahagún Ribera, *Coloquios de los doce*, Italian edition, Palermo 1991.

David Sànchez, 'Hernán Cortés y la enfermedad del oro', *Revista de Humanidades y Ciencias Sociales*, no. 10, 6 (2019), pp. 36–45.

Nicolás Sánchez-Albornoz, *La población de América Latina*, Mexico City, 2014 (1st ed. Buenos Aires, 1968).

Ricardo Sanchez Hernández, 'La geología del jade mesoamericano', *Arqueología Mexicana*, no. 133 (2015), pp. 37–41.

Durdica Ségota, 'El panteón mexica', *Arqueología Mexicana*, no. 15 (1995), pp. 32–41.

Sguardi Globali. Mappe olandesi, spagnole e portoghesi nelle collezioni del Granduca Cosimo III de' Medici, catalogue of the exhibition (Florence, Biblioteca Medicea Laurenziana, 2019–20), edited by Angelo Cattaneo and Sabrina Corbellini, Florence, 2019.

Damien Simonis, *Spagna centrale e meridionale*, Turin, 2000.

Damien Simonis, *Spagna settentrionale*, Turin, 2000.

Alexandra Strong Zuñiga, 'Riqueza, prestigio y ofrenda divina: los usos del cacao en el mundo nahua', *Hospitalidad ESDAI*, no. 37 (2020), pp. 35–53.

Francesco Surdich, *Verso il Nuovo Mondo. L'immaginario europeo e la scoperta dell'America*, Florence, 2002.

Francesco Surdich, *Verso il nuovo mondo. La dimensione e la coscienza delle scoperte*, Florence, 1991.

Valentine Tibère, 'El chocolate del Alba', *Artes de México*, no. 103 (2011), pp. 12–25.

Tzvetan Todorov, *La conquête de l'Amérique. La question de l'autre*, Paris, 1982.

José Toribio Medina, *Historia de la imprenta en los antiguos dominios españoles de América y Oceanía*, 2 vols., Santiago, 1958.

Marta Turok, 'Jicaras, cacao y quetzal', *Artes de México*, no. 105 (2011), pp. 40–5.

Gustavo Uzielli, *Paolo dal Pozzo Toscanelli, iniziatore della scoperta d'America. Ricordo del solstizio d'estate del 1892*, Florence, 1892.

Giuseppe Vaccarini, *Cioccolato. Codex Nero Fondente*, Milan, 2018.

Le vie della storia nell'economia, edited by Pierluigi Ciocca, Bologna, 2003.

Luis M. Villar Anléu, *Apuntes sobre la historia natural del Quetzal*, Guatemala, 1983.

Luis Villoro, 'Sahagún o los límites del descubrimiento del otro', *Estudios de cultura Náhuatl*, no. 29 (1999), pp. 15–26.

Voltaire, *Jeannot et Colin*, Paris, 1764.

Olivia B. Waxman, 'The First Africans in Virginia Landed in 1619. It Was a Turning Point for Slavery in American History. But Not the Beginning', *The Times*, 20 August 2019.

Phil C. Weigand, 'La Turquesa', *Arqueología Mexicana*, no. 27 (1997), pp. 26–33.

The World that Wasn't There. Pre-Columbian Art in the Ligabue Collection, catalogue of the exhibition (Florence, Museo Archeologico Nazionale, Palazzo della Crocetta, 2015–16), edited by Adriano Favaro, Milan, 2015.

Sonia Zarrillo et al., 'The Use and Domestication of Theobroma Cacao during the Mid-Holocene in the Upper Amazon', *Nature Ecology Evolution Journal*, 2 (2018), pp. 1879–88.

Digital sources (last access 9 January 2024)

Accademia Della Crusca, Consulenza: 'Cioccolata o Cioccolato?', <https://accademiadellacrusca.it/it/consulenza/cioccolata-o-cioccolato/212>.

Manuel Aguilar Moreno, 'La aventuresca vida del Códice Florentino de Bernardino de Sahagún', 17 May 2018, <https://www.youtube.com/watch?v=82LWwzY_hIE>.

Stefano Carnazzi, 'Schiavitù, ieri e oggi. Una parola antica, un'omertà moderna', *LifeGate Daily, Italia*, 1st December 2016, <https://www.lifegate.it/schiavitu-moderna#:~:text=La%20storia%20dei%20prodotti%20coloniali,Atlantico%20furono%20circa%2012%20milioni>.

Codice Mendoza 1540, Mèxico 2020, <www.codicemendoza.inah.gob.mx/inicio.php>.

'Culturas mesoamericanas', *CONCEPTO Enciclopedia*, <https://concepto.de/culturas-mesoamericanas/>.

The Dark Side of Chocolate, documentary, Denmark, 2010, directed by Miki Mistrati and U. Roberto Romano, <https://shorturl.at/gqNU7>.

Giovanni Donadelli, Lorena Rocca, 'La cartografia digitale nella scuola. Il riscatto della geografia', 'DIDAMATICA 2012. L'informatica per la didattica', symposium (Taranto, 14–16 May 2012), <https://mondodigitale.aicanet.net/2012-2/didamatica/PAPER/FULL/F157.pdf>.

I Georgofili. Atti dell'Accademia dei Georgofili, 8th series, 18 (2021), <https://www.georgofili.net/articoli/atti-dei-georgofili-2021/7879>.

International Cocoa Organization, <http://www.icco.org/>.

International Rights Advocates, <https://www.internationalrightsadvocates.org/cases>.

Paolo Jedlowski, *Sulla memoria storica*, 2020, <https://scienzepolitiche.unical.it/bacheca/archivio/materiale/26/Studi%20culturali%202020-21/Dispensa%20n.%202/Sulla%20memoria%20storica.pdf>.

Miguel León-Portilla, *Sahagún antropológico. Su aportación cuestionada*, 31 December 1999, <https://letraslibres.com/revista-mexico/sa-

hagun-antropologico-su-aportacion-cuestionada/>.

Bertrand Lobjois, '¿El regreso de Quetzalcóatl, mito o realidad?', <https://www.udem.edu.mx/sites/default/files/2021-07/CCC-Resena-Conferencia-2017-Lobjois.pdf/>.

Giovanni Francesco Lucarelli, 'Il passato che è in noi: beni culturali e "sostenibilità" della memoria', *Aperta Contrada*, 30 December 2010, <https://www.apertacontrada.it/2010/12/30/il-passato-che-e-in-noi-beni-culturali-e-sostenibilita-della-memoria/>.

Joshua J. Mark, *La schiavizzazione dei Nativi nell'America coloniale*, Italian translation by Alfonso Vincenzo Mauro, *World History Encyclopedia*, <https://www.worldhistory.org/trans/it/2-1742/la-schiavizzazione-dei-nativi-nellamerica-colonial/>.

Ministero della Cultura, 'La ricetta segretissima del cioccolatte al gelsomino del Granduca di Toscana sul canale youtube del MIBACT per la campagna "La Cultura non si ferma"', <https://www.beniculturali.it/comunicato/la-ricetta-segretissima-del-cioccolatte-al-gelsomino-del-granduca-di-toscana-sul-canale-youtube-del-mibact-per-la-campagna-la-cultura-non-si-ferma>.

'Obsidiana, la primitiva arma de los aztecas', *History National Geographic*, 11 December 2019, <https://historia.nationalgeographic.com.es/a/obsidiana-primitiva-arma-aztecas_13990>.

Michel Olguín Lacunza, 'Cinco curiosidades de los colibríes', *UNAM Revista*, 2021, < https://unamglobal.unam.mx/global_revista/cinco-curiosidades-de-los-colibries/>.

Alejandra Ortiz, 'El Códice Florentino ya puede consultarse con un clic', *La Jornada*, 6 February 2013, <https://www.jornada.com.mx/2013/02/06/cultura/a04n1cul>.

Gabriel Paquette, 'Il traffico umano attraverso l'Atlantico', *Storica National Geographic*, 5 January 2022, <https://www.storicang.it/a/il-traffico-umano-attraverso-latlantico_15423>.

'Francisco del Paso y Troncoso en Europa', *Diario de Historias. Crónicas de viaje por archivos y bibliotecas*, no. 6, 2 (2014), <https://silo.tips/download/troncoso-se-localiza-en-el-archivo-historico-de-la-biblioteca-del-museo-nacional>.

'La rivoluzione dimenticata', *viaAriosto - Humanas actiones intelligere*, 12 December 2013, <https://viaariosto.wordpress.com/2013/12/12/la-rivoluzione-dimenticata/>.

Bernardino de Sahagún, *Historia general de las cosas de Nueva España* [1577], Biblioteca Medicea Laurenziana, Teca Digitale, Mediceo Palatino, 2023, <https://tecabml.contentdm.oclc.org/digital/collection/med-palatino/search>.

Bernardino de Sahagún, *Historia general de las cosas de Nueva España* [1577], with the title *Digital Florentine Codex. An Encyclopedia of 16th-Century Indigenous México*, 2023, <https://florentinecodex.getty.edu/>.

Bernardino de Sahagún, *Historia general de las cosas de Nueva España* [1577], The Library of Congress, <https://www.loc.gov/item/2021667837>.

Francesca Scaroli, 'Cacao, comida de los dioses', blog *Alas Para Volar. Arte e cultura del Messico*, 31 May 2016, <https://alaspavolarsite.wordpress.com/2016/05/31/cacao-comida-de-los-dioses/>.

Slave Free Chocolate Organization, <https://www.slavefreechocolate.org/>.

Jonathan Stempel, 'Hershey, Nestle, Cargill Win Dismissal in U.S. of Child Slavery Lawsuit', Reuters, 29 June 2022, <https://www.reuters.com/business/hershey-nestle-cargill-win-dismissal-us-child-slavery-lawsuit-2022-06-28/>.

'Trata de esclavos; Juicio de San Juan Pablo II', *Pontificium*, <https://www.dhial.org/diccionario/index.php?title=TRATA_DE_ESCLAVOS;_Juicio_de_San_Juan_Pablo_II>.

World Atlas, <http://www.sfu.ca/geog351fall03>.

Phersus Atlas
<https://content.phersu-atlas.com/data/event_changes/4451>.
<https://content.phersu-atlas.com/data/comparison_polity/3800>.
<https://content.phersu-atlas.com/data/annual_density_map_polity/3800/1491>.
<https://content.phersu-atlas.com/data/annual_density_map_polity/3800/1520>.
<https://content.phersu-atlas.com/data/cluster_war_summary/4451/1521/8/969/north_america>.
<https://content.phersu-atlas.com/data/polity_war/4451>.
<https://content.phersu-atlas.com/data/max_polity/3800>.
<https://content.phersu-atlas.com/data/max_polity_war/6781/4451>.
<https://content.phersu-atlas.com/object/event/4451>.
<https://content.phersu-atlas.com/object/event/520>.

Table of contents

© 2024 Mandragora.
All rights reserved.
Mandragora s.r.l.
via Capo di Mondo 61, 50136 Firenze
www.mandragora.it

Editor
Marco Salucci

Art director
Paola Vannucchi

English translation
NTL, Florence

Pre-press
Puntoeacapo, Florence

Printed by
Grafiche Martinelli, Bagno a Ripoli (Florence)

Bound by
Legatoria Giagnoni (Florence)

isbn 978-88-7461-684-8

January 2024

Photographs
Florence, Biblioteca Medicea Laurenziana*; Florence, Gallerie degli Uffizi*; Florence, Museo Archeologico Nazionale*; Florence, Museo Galileo, Istituto e Museo di Storia della Scienza; London, Wellcome Library; London, © The Trustees of the British Museum; Los Angeles, The Getty Research Institute, 2698-167.v6, gift of the Getty Research Institute Collections Council; Louis Agassiz's Slave Daguerreo-types; Rome, Museo delle Civiltà*; Pixabay, Regional Digitilisation Lab-MIK-Virtual Malopolska project; Washington, D.C., International Rights Advocates, Slave Free Chocolate; Washington, D.C., Library of Congress, Geography and Map Division, Photocrom Print Collection 1890.
Photographers: Soraya Al Azzam, Alex Arcos, Kamran Aydinov, Laura Carrillo Márquez, Rebeca Cruz Galvan, Google Arts and Culture, Gr4yleaf, Nasser Halaweh, Tom Hollyman (1919–2009), Alejandro Linares Garcia, Rafael Gutiérrez, Lorenzo Mennonna, @Scott.Zona, Mario Wong.

Special thanks for their kind concessions: Antica Dolceria Bonajuto, Maria Gloria Aragona Pignatelli Cortés (1927–2015), Roberto Caraceni, Gelateria della Passera, Getty's Open Content Program, International Rights Advocates, Karuna, Secretaria de Turismo de Tabasco, Paolo Staccoli, Nicolò Vellino, Vestri Cioccolato.

* by permission of the Ministero della Cultura. Any further reproduction by any means is strictly prohibited.